ANNALS *of* THE NEW YORK ACADEMY OF SCIENCES

VOLUME
1278

ISBN-10: 1-57331-893-0; **ISBN-13:** 978-1-57331-893-8

ISSUE

Annals Meeting Reports

TABLE OF CONTENTS

EDITOR-IN-CHIEF
Douglas Braaten

ASSOCIATE EDITOR
David Alvaro

PROJECT MANAGER
Steven E. Bohall

DESIGN
Ash Ayman Shairzay

The New York Academy of Sciences
7 World Trade Center
250 Greenwich Street, 40th Floor
New York, NY 10007-2157
annals@nyas.org
www.nyas.org/annals

Annals of the New York Academy of Sciences (ISSN: 0077-8923 [print]; ISSN: 1749-6632 [online]) is published 30 times a year on behalf of the New York Academy of Sciences by Wiley Subscription Services, Inc., a Wiley Company, 111 River Street, Hoboken, NJ 07030-5774.

Mailing: *Annals of the New York Academy of Sciences* is mailed standard rate.

Postmaster: Send all address changes to ANNALS OF THE NEW YORK ACADEMY OF SCIENCES, Journal Customer Services, John Wiley & Sons Inc., 350 Main Street, Malden, MA 02148-5020.

Disclaimer: The publisher, the New York Academy of Sciences, and the editors cannot be held responsible for errors or any consequences arising from the use of information contained in this publication; the views and opinions expressed do not necessarily reflect those of the publisher, the New York Academy of Sciences, and editors, neither does the publication of advertisements constitute any endorsement by the publisher, the New York Academy of Sciences and editors of the products advertised.

Publisher: *Annals of the New York Academy of Sciences* is published by Wiley Periodicals, Inc., Commerce Place, 350 Main Street, Malden, MA 02148; Telephone: 781 388 8200; Fax: 781 388 8210.

Journal Customer Services: For ordering information, claims, and any inquiry concerning your subscription, please go to www.wileycustomerhelp.com/ask or contact your nearest office. *Americas:* Email: cs-journals@wiley.com; Tel:+1 781 388 8598 or 1 800 835 6770 (Toll free in the USA & Canada). *Europe, Middle East, Asia:* Email: cs-journals@wiley. com; Tel: +44 (0) 1865 778315. *Asia Pacific:* Email: cs-journals@wiley.com; Tel: +65 6511 8000. *Japan:* For Japanese speaking support, Email: cs-japan@wiley.com; Tel: +65 6511 8010 or Tel (toll-free): 005 316 50 480. Visit our Online Customer Get-Help available in 6 languages at www.wileycustomerhelp.com.

Information for Subscribers: *Annals of the New York Academy of Sciences* is published in 30 volumes per year. Subscription prices for 2013 are: Print & Online: US$6,053 (US), US$6,589 (Rest of World), €4,269 (Europe), £3,364 (UK). Prices are exclusive of tax. Australian GST, Canadian GST, and European VAT will be applied at the appropriate rates. For more information on current tax rates, please go to www.wileyonlinelibrary.com/tax-vat. The price includes online access to the current and all online back files to January 1, 2009, where available. For other pricing options, including access information and terms and conditions, please visit www.wileyonlinelibrary.com/access.

Delivery Terms and Legal Title: Where the subscription price includes print volumes and delivery is to the recipient's address, delivery terms are Delivered at Place (DAP); the recipient is responsible for paying any import duty or taxes. Title to all volumes transfers FOB our shipping point, freight prepaid. We will endeavour to fulfill claims for missing or damaged copies within six months of publication, within our reasonable discretion and subject to availability.

Back issues: Recent single volumes are available to institutions at the current single volume price from cs-journals@wiley.com. Earlier volumes may be obtained from Periodicals Service Company, 11 Main Street, Germantown, NY 12526, USA. Tel: +1 518 537 4700, Fax: +1 518 537 5899, Email: psc@periodicals.com. For submission instructions, subscription, and all other information visit: www.wileyonlinelibrary.com/journal/nyas.

Production Editors: Kelly McSweeney and Allie Struzik (email: nyas@wiley.com).

Commercial Reprints: Dan Nicholas (email: dnicholas@wiley.com).

Membership information: Members may order copies of *Annals* volumes directly from the Academy by visiting www. nyas.org/annals, emailing customerservice@nyas.org, faxing +1 212 298 3650, or calling 1 800 843 6927 (toll free in the USA), or +1 212 298 8640. For more information on becoming a member of the New York Academy of Sciences, please visit www.nyas.org/membership. Claims and inquiries on member orders should be directed to the Academy at email: membership@nyas.org or Tel: 1 800 843 6927 (toll free in the USA) or +1 212 298 8640.

Printed in the USA by The Sheridan Group.

View *Annals* online at www.wileyonlinelibrary.com/journal/nyas.

Abstracting and Indexing Services: *Annals of the New York Academy of Sciences* is indexed by MEDLINE, Science Citation Index, and SCOPUS. For a complete list of A&I services, please visit the journal homepage at www. wileyonlinelibrary.com/journal/nyas.

Access to *Annals* is available free online within institutions in the developing world through the AGORA initiative with the FAO, the HINARI initiative with the WHO, and the OARE initiative with UNEP. For information, visit www. aginternetwork.org, www.healthinternetwork.org, www.oarescience.org.

Annals of the New York Academy of Sciences accepts articles for Open Access publication. Please visit http://olabout.wiley.com/WileyCDA/Section/id-406241.html for further information about OnlineOpen.

Wiley's Corporate Citizenship initiative seeks to address the environmental, social, economic, and ethical challenges faced in our business and which are important to our diverse stakeholder groups. Since launching the initiative, we have focused on sharing our content with those in need, enhancing community philanthropy, reducing our carbon impact, creating global guidelines and best practices for paper use, establishing a vendor code of ethics, and engaging our colleagues and other stakeholders in our efforts. Follow our progress at www.wiley.com/go/citizenship.

Ann. N.Y. Acad. Sci ISSN 0077-8923

ANNALS OF THE NEW YORK ACADEMY OF SCIENCES
Issue: Annals *Meeting Reports*

Biomarkers in nutrition: new frontiers in research and application

Gerald F. Combs Jr.,[1] Paula R. Trumbo,[2] Michelle C. McKinley,[3] John Milner,[4] Stephanie Studenski,[5] Takeshi Kimura,[6] Steven M. Watkins,[7] and Daniel J. Raiten[8]

[1]Grand Forks Human Nutrition Research Center, U.S. Department of Agriculture, Agricultural Research Service, Grand Forks, North Dakota. [2]Center for Food Safety and Applied Nutrition, U.S. Food and Drug Administration, Rockville, Maryland. [3]School of Medicine, Dentistry and Biomedical Sciences, Queen's University Belfast, Belfast, Northern Ireland, United Kingdom. [4]National Cancer Institute, National Institutes of Health, U.S. Department of Health and Human Services, Rockville, Maryland. [5]University of Pittsburgh, Pittsburgh, Pennsylvania. [6]Ajinomoto Co., Inc., Tokyo, Japan. [7]Tethys Bioscience, Inc., Emeryville, Dalifornia. [8]Eunice Kennedy Shriver National Institute of Child Health and Human Development, National Institutes of Health, U.S. Department of Health and Human Services, Rockville, Maryland

Address for correspondence: Gerald F. Combs, Grand Forks Human Nutrition Research Center, U.S. Department of Agriculture, Agricultural Research Service, Grand Forks, North Dakota. annals@nyas.org

Nutritional biomarkers—biochemical, functional, or clinical indices of nutrient intake, status, or functional effects—are needed to support evidence-based clinical guidance and effective health programs and policies related to food, nutrition, and health. Such indices can reveal information about biological or physiological responses to dietary behavior or pathogenic processes, and can be used to monitor responses to therapeutic interventions and to provide information on interindividual differences in response to diet and nutrition. Many nutritional biomarkers are available; yet there has been no formal mechanism to establish consensus regarding the optimal biomarkers for particular nutrients and applications.

Keywords: biomarkers; nutrition; nutrients

Sponsored by the Sackler Institute for Nutrition Science and the New York Academy of Sciences, the conference "Biomarkers in Nutrition: New Frontiers in Research and Application" was held on April 18, 2012 at the New York Academy of Sciences in New York City. The meeting, comprising individual talks and group discussions, brought together scientists and practitioners from industry, academia, and governmental organizations to discuss the current state of knowledge about nutritional biomarkers, to identify important challenges and unanswered questions, and to catalyze new research toward the common goal of implementing nutritional biomarkers in a broad, cost-effective, and meaningful way.

The Biomarkers of Nutrition for Development (BOND) program

Gerald F. Combs, Jr. (USDA Grand Forks Human Nutrition Research Center) spoke on behalf of Daniel J. Raiten (Eunice Kennedy Shriver Na-

tional Institute of Child Health and Human Development, NICHD) about the BOND Program. Supported by the Bill and Melinda Gates Foundation and managed by NICHD, BOND aims to harmonize the discovery, development, and distribution of biomarkers of nutritional status and to provide advice to researchers, clinicians, and policymakers on how best to use nutrition biomarkers.

Food and nutrition play key roles in supporting health and preventing disease. Globally, maternal and child undernutrition results in some 3.5 million deaths annually, accounting for 35% of the disease burden in children under five years of age.[1] Undernutrition includes what has been called *hidden hunger*, that is, single and multiple micronutrient insufficiencies that affect two billion individuals in both industrialized and developing countries,[2] and in those overweight or underweight.[3] At the same time, overweight and obesity are becoming more prevalent, with an estimated one billion adults and

22 million children being overweight.[4] Thus, the dual burden of over- and undernutrition presents a major challenge.[5]

It has been noted that the ability to assess the health impacts of nutritional status depends on the availability of accurate and reliable biomarkers that reflect nutrient exposure, status, and effect.[1] Biomarkers are essential in this regard; yet, confusion remains surrounding their use and application. What might be a useful index of nutrient exposure may not necessarily reflect nutrient status, which, in turn, may not necessarily reflect the impact or function of that nutrient. Systematic reviews of a range of nutritional biomarkers have emphasized the lack of clarity in the definition of biomarkers and their application and purpose.[6] The usefulness of even the most well-documented biomarkers has been limited by gaps in the understanding of their physiologic significance.

The BOND program was created to address this need; it is supported by a consortium that includes the Bill and Melinda Gates Foundation (BMGF), PepsiCo, the NIH Office of Dietary Supplements, and the NIH Division of Nutrition Research Coordination, and includes memberships with organizations and agencies representing the breadth of the global food and nutrition community. BOND is managed by the NICHD and aims to harmonize the process of making decisions about the best uses of biomarkers in individual situations.

BOND has targeted four primary user communities for its translational activities:

1. research (including basic research examining the role of nutrition in biological systems, clinical research, and operations research);
2. clinical care;
3. programs (surveillance to identify populations at risk, and monitoring and evaluation of public health programs); and,
4. policy (evaluation of the evidence base to make national or global policy about diet and health, and funding agencies that make decisions about priorities in food and nutrition).

Biomarker needs are, therefore, both general and user specific.

The BOND program was initiated through a consultative process with the food and nutrition community that culminated in an organizational conference held in Vienna in 2010, organized by NICHD and hosted by the International Atomic Energy Agency. Partners included key multilateral U.S. agencies and public and private organizations. That assembly endorsed the need to develop a process to inform the community about the relative strengths/weaknesses and specific applications of various biomarkers under defined conditions. Specific attention was paid to the needs for nutritional biomarkers in four use areas: research, clinical, policy, and programs. Five micronutrients of public health importance (iron, zinc, vitamin A, folate, and vitamin B_{12}) were discussed as case studies with respect to new frontiers in science and technology. An overview of that meeting was published.[7]

The mission of BOND was developed in the Vienna meeting and included (1) developing consensus on accurate assessment methodologies relevant to users domestically and internationally, and (2) providing evidence-based advice to support a range of activities of the entire food/nutrition research and global health enterprise including (a) further development of national nutrition surveys, (b) review and development of dietary guidance, (c) development of new and improved systems of food/nutrient delivery, (d) monitoring and evaluation of new and existing programs and interventions, and (e) basic and clinical research to generate new data on diet and disease relationships and the roles of nutrients in promoting health and preventing disease.

BOND is implementing this mission through two approaches: a translational track involving partnering with U.S. and international agencies, and the development of a research agenda that will lead to funding opportunities supported by agencies and organizations across the breadth of the global research funding enterprise (Fig. 1).

approach to addressing the role of
Nutrition in global maternal and child health

Figure 1. The Biomarkers of Nutrition and Development (BOND) program.

After the Vienna meeting, NICHD received core funding from the BMGF to begin the BOND project. The initial phase of that project includes the establishment of an expert panel for each of the above-mentioned case nutrients plus iodine. These panels are charged with reviewing the relevant literature supporting decision points regarding specific biomarkers, the needs of specific user groups, opportunities for new technologies, and key knowledge gaps. In 2013, BOND will convene a meeting at which the panels will provide input for the development of queries and responses that reflect the primary user communities.

Combs concluded by pointing out that BOND has established a website (http://www.nichd.nih.gov/global˙nutrition/programs/bond/) with links to member agencies and organizations, opportunities to provide input, content on biomarkers relative to specific nutrients, and overviews of cross-cutting issues relative to nutritional biomarkers. Ultimately, the website will house a query-based system enabling users to gain information on particular biomarker applications.

Use of biomarkers to substantiate health claims

Paula Trumbo (U.S. Food and Drug Administration, (FDA)) described the process by which the FDA evaluates the scientific evidence for the use of biomarkers to substantiate health claims on the labels of foods and dietary supplements. She pointed out that, as part of its evidence-based systemic review, the FDA relies on surrogate endpoints (qualified risk biomarkers) in the premarket scientific review of health claims used for labeling foods and dietary supplements. Health claims provide information about the relationship between a food or food component and risk of a disease or health-related condition (e.g., a surrogate endpoint of disease risk). The review of scientific evidence involves the identification, classification, and rating of relevant studies; the evaluation of the strength of the evidence; and the determination of whether that evidence supports a health claim.

The FDA relies on a limited number of available surrogate endpoints for its health claim reviews. For this reason, the FDA funded the Institute of Medicine (IOM) to develop a framework for the qualification of risk biomarkers. That report[8] emphasizes the need for validated analytical methods that can measure risk biomarkers. The recommended qualification framework also calls for evaluating the relationship between the risk biomarker and the clinical endpoint (Fig. 2), as well as the need for evidence that the intervention affecting the risk biomarker also influences the clinical endpoint. This report is being considered by the FDA for future risk biomarker qualification.

Trumbo concluded by pointing out that the FDA Center for Drug Evaluation and Research manages a biomarker qualification program that includes the review of biomarkers of chronic disease risk. Such a program could possibly evaluate biomarkers that are applicable to health claims.

Biomarkers of selenium status

Gerald Combs reviewed the use and interpretation of biomarkers of selenium (Se) status in light of the current understanding of Se metabolism. Status with respect to the essential nutrient Se is considered under four categories relevant to human nutrition and health: (1) assessment of Se intake/exposure, (2) assessment of risk of nutritional Se deficiency, (3) assessment of Se adequacy for cancer risk reduction, and (4) assessment of risk of Se toxicity (Fig. 3). He pointed out that each category relies on a different set of endpoints with different evidence bases.

The nutritional functions of Se appear to be discharged by a group of selenoproteins, the best characterized of which (glutathione peroxidases, selenoprotein P) can serve as biomarkers of Se status. Some 25 selenoproteins have been identified; each can incorporate Se from inorganic Se compounds (selenite, selenate) and the amino acid analog selenocysteine (SeCys), but not from the dominant

Health Claims characterize risk-reduction relationships between diet and disease or health-related condition

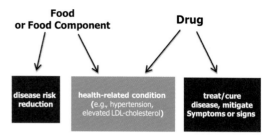

Figure 2. Risk reduction relationships implied by FDA-approved health claims.

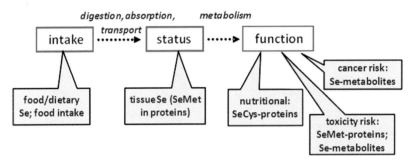

Figure 3. Types of biomarkers available for assessing Se intake, status, and function.

food form selenomethionine (SeMet), although the latter can freely replace methionine in protein biosynthesis.

This means that Se intake and exposure can be assessed on the basis of the Se contents of accessible specimens (e.g., plasma, urine, hair/nails, and buccal cells) if the dominant form of ingested Se is known. The Se contents of these tissues reflect the rate of intake of inorganic forms or SeCys only to the point of maximal selenoprotein expression, which occurs with approximately 40–50 µg Se per day or plasma Se levels of 70–80 ng/mL. If, however, Se is consumed in the form of SeMet, tissue levels can increase over a virtually unlimited range, reflecting the nonspecific incorporation of the element into proteins.[9, 10] While such cases can be assessed for Se exposure, they offer no information about function.

The potential for Se to reduce cancer risk is being actively researched. Current evidence suggests that risk reduction occurs among those with nutritionally adequate (i.e., maximal selenoprotein expression), but not high, Se status. For example, the Nutritional Prevention of Cancer Trial[11] found that supplemental Se reduced cancer risk in non-deficient Americans with baseline plasma Se levels <120 ng/mL. This conclusion is consistent with the results of the larger SELECT trial.[12] Thus, it appears that candidates for Se protection against cancer can be identified using Se biomarkers similar to those used to assess Se exposure, but with the application of a great target plasma level.

The potential for very high Se status to produce adverse physiological effects has been established from animal studies and accidental exposure on humans; these physiological effects have produced an array of clinical indicators but few biomarkers with predictive potential. For this reason, the default choice has been to use the highest Se tissue levels

observed with no adverse effects as risk indicators. Some studies show no adverse effects with plasma Se levels <1000 ng/mL; however, recent studies have suggested that plasma levels >140 ng/mL may increase type 2 diabetes risk.[13, 14]

Combs concluded by reiterating the two questions dominating the consideration of health roles of Se: Who may benefit from increasing Se intake? and Who may be at risk from increasing Se intake? He pointed to the need for better biomarkers of Se function to address each of these questions.

Rationale and process for developing biomarkers for sarcopenia

Stephanie Studenski (University of Pittsburgh) discussed the use of biomarkers in the study of sarcopenia. Losses of muscle mass and strength seem to be almost universal age-related phenomena and are associated in epidemiological studies with numerous adverse outcomes including disability and mortality.[15] She pointed out that, in order to develop useful biomarkers, such conditions must be clearly defined. While several definitions have been proposed for sarcopenia, which generally indicates loss of muscle mass, challenges remain. First, the criteria for low muscle mass were initially developed using sample population distributions and without considering the effect of strength. Second, more recently proposed definitions have been based only on expert opinion without a formal evidence base.

In order to address these challenges, the Biomakers Consortium was funded through the Foundation for the National Institutes of Health, with the goal of pooling data related to body composition, muscle strength, functional abilities, and other relevant factors from multiple longitudinal and clinical trial studies of older adults. The model used by the consortium, as well as prior expert panels, suggests that the underlying clinical process proceeds from abnormal muscle mass or quality to muscle weakness, which results in reduced physical function and disability. In order to apply this model within a clinical diagnostic framework, the consortium suggested that older people would present clinically with complaints of reduced functional abilities that could then be assessed objectively by clinicians using physical performance tests, such as walking speed or ability to rise from a chair. Accordingly, the specific aims of the pooled analyses were to (1) determine criteria for clinically important weakness based on optimal discrimination between older persons with and without reduced physical performance; (2) determine criteria for clinically important low muscle mass based on the optimal discrimination between older persons with and without criteria for clinically important weakness; and (3) assess longitudinally whether the criteria defined in the first two aims help predict the onset of future physical disability.

The final pooled sample of older adults encompasses over 30,000 individuals of diverse age, gender, ethnicity, function, strength, and body composition. All major analyses have been completed and were presented at a conference in May 2012 to an audience of clinicians, regulators, scientists, and representatives from the private sector. There was general agreement that a clinical definition must include both weakness and low muscle mass, and that physical performance measures were the optimal choice for primary outcome measures because they are objective, reliable, and clinically relevant markers of function. There was also general agreement that older persons may be weak for reasons other than low muscle mass, because the muscle is not producing adequate force due to factors either within or outside the muscle itself. Therefore, it was suggested that poor muscle quality be used to define persons who are weak but who do not have low muscle mass.

In order to further evaluate the causes and treatment of poor muscle quality, the key next step is to develop criteria for inadequate force production per unit of muscle mass. Potential contributors to poor muscle quality might include factors related to muscle composition (including proteins and lipids), cellular energetics, and neuromuscular control.[16]

The development of biomarkers for sarcopenia is of particular value in the examination of mechanisms, diagnosis, and responses to treatment. A

- **Functional:** physical performance, muscle strength

- **Imaging:** DXA, CT, MRI, echography, electrical impedence myography

- **Biological:**
  directly related to muscle
      creatinine excretion, P3NP (procollagen type II N-terminal peptide)
  indirectly related (for assessing mechanism)
      markers of inflammation, oxidative damage, protein synthesis, hormone levels

Figure 4. Candidate biomarkers of sarcopenia.

biomarker can be considered a characteristic that is objectively measured and evaluated as an indicator of normal biological processes, pathogenic processes, or pharmacologic responses to a therapeutic intervention. Biomarkers for sarcopenia might be grouped into three main types—functional, imaging, and biological—and could be used for screening, diagnosis, or endpoint assessment (Fig. 4). Among functional measures, objective indicators of physical performance might be useful for all three purposes. Measures of body composition, including muscle and fat, can be obtained from DXA, CT, MRI, or other techniques. While DXA is considered to be more widely available clinically and to have minimal respondent burden, there are substantial concerns about its ability to account for fat in muscle, while MRI and CT are superior for such detailed assessment of muscle but are more expensive and somewhat more burdensome. Therefore, some investigators have proposed using DXA for screening and CT or MRI for baseline and endpoint assessment. There are also newer techniques in development that are based on echomyography or electrical impedence myography. Most imaging techniques suffer to some extent from difficulty discriminating lean mass from water, so that some degree of error will be present in older persons with edema or other instances of excess body water.

Studenski concluded by stating that new biological markers are needed to assess aspects of muscle metabolism directly or to assess other factors that influence muscle synthesis and degradation. Candidates of the former type include serum or urinary creatinine, creatine phosphokinase, or P3NP (procollagen type II N-terminal peptide). Those of the latter type include indicators of inflammation, oxidative damage, hormone levels, or protein

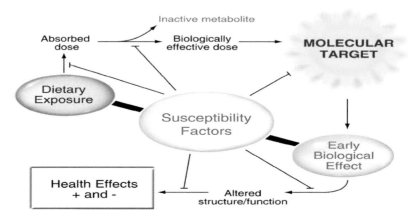

Figure 5. Need for biomarkers of host susceptibility in assessing the health impacts of bioactive substances.

synthesis.[17] She noted that studies to assess the test characteristics of such biological indicators for sarcopenia are being implemented and that clinical trials for the treatment or prevention of sarcopenia, based on exercise, nutrition, and pharmacological approaches, are currently in the field or are being designed.

Markers of dietary intake and risk of cancer

John Milner (National Cancer Institute (MCI) and National Institutes of Health (NIH)) described the myriad challenges and considerations involved in the development of sensitive and reliable biomarkers of dietary intake and cancer risk. He pointed out the increasing concern about noncommunicable diseases, especially cancer and heart disease, which centers on the lost productivity of individuals and the projected massive medical expense. It is abundantly clear that the increase in risk of cancer as well as heart disease is occurring in both developed and developing parts of the world. While part of the increase surely relates to an aging society and to the increased incidence of obesity, it is also becoming increasingly apparent that nutritional inadequacies are also contributors.

In order to evaluate the benefits of foods for health promotion and disease prevention, there is recognition that three types of biomarkers will be needed for an appropriate assessment. The first is a biomarker that evaluates intake or exposure. Unfortunately, the present methodologies are woefully inadequate and have large error terms, and thus may account for part of the wide variation in risk of noncommuni-

cable diseases observed nationally and internationally. Not all fruits and vegetables are identical, and thus there is a need to identify which foods offer the greatest benefits for health promotion.

The second class of biomarker relates to the site of action (molecular target). Unfortunately, there is likely more than one target for bioactive food components, and thus unravelling the specific site of action of a bioactive food component is extremely challenging. This area is also complicated since normal and aberrant cell metabolism can influence the specific targets and thereby influence the response to bioactive food components. Since several foods may influence the same target, knowledge about dietary patterns will require increasing attention to explain the means to achieving maximum health benefits from the food and supplements that are consumed.

The third type of biomarker encompasses the interactions food components exhibit either among each other or with a person's genomic profile (Fig. 5). It is becoming increasingly recognized that genetics (polymorphisms, deletions, insertions, and copy number) and epigenetics (DNA methylation, histone homeostasis, noncoding RNA, and polycomb organization), along with transcriptome regulation, are critical to the response to foods and their components. These genomic, or more precisely nutrigenomic, factors influence multiple cellular processes that are reflected by shifts in specific proteins and cellular metabolites. Thus, proteomics and metabolomics are important technologies for assessing the impact of foods and their constituents at the phenotypic level.

Even these three biomarker categories are influenced by a host of internal and external insults that influence the needs for specific food components. Notable among these are the influence of excess calories (obesity), viruses, bacteria, and environmental contaminants. Thus, biomarkers are also needed to not only assess the omics of normal nutrition, but also the insults that can influence requirements under abnormal or diseased conditions. Finally, it must be recognized that many bioactive food components have the same site of action, and thus a much greater understanding of these interactions are needed to truly assess the benefits or risks in changing a person's diet.

Assessments of dietary intake and risk of cancer must also take into account the potential deleterious effects of modifying intakes, especially overindulgence. Increasing evidence points to harmful effects of consuming excessive amounts of specific foods or specific dietary supplements. These ill consequences can manifest themselves in multiple ways, including possible increases in cancer risk. It is unclear why this increased risk occurs. However, it may reflect the ability of excess amounts of specific bioactive food constituents to influence a metabolic pathway that is not normally influenced, or to create new and possibly deleterious metabolites. Either way, there is a need to identify individuals or subpopulations vulnerable to certain foods and their constituents. This is becoming a greater concern, given the large percentage of the population that consumes dietary supplements with the increased likelihood of excessive exposure.

Milner concluded that it is becoming increasingly clear that phenotypic changes can no longer be relied upon to identify benefits and risks associated with consuming specific foods or their components. The newest omic technologies must be embraced to identify those individuals who will benefit most from, or be placed at risk as a result of, dietary change. For this reason, he noted that intersectoral partnerships that build on academic, industrial, and governmental research strengths are critical for advancing knowledge about the relationships between diet and health.

Blood-based markers of nutrient status and their association with metabolic risk

Steven Watkins (Tethys Bioscience, Inc.) discussed the fact that nutritional approaches can often be difficult to employ as prevention strategies in primary care settings, and that it is hard to argue that we are winning the battle against chronic disease. Yet, many large clinical studies have demonstrated that there is a profound effect of nutrients and lifestyle on the prevention of metabolic and cardiovascular disease.

One strategy for improving compliance with nutrition guidance is the enhancing of education for doctors and patients. Watkins suggested, however, that there may be another strategy more suited for use in a medical setting. Medicine in general is geared toward a treat-to-target approach, where diagnostics provide information on a patient's current status and whether corrective actions have made a positive impact. Nutrition is not enabled in this way, in that there is no objective score that can be used by clinicians as part of the treatment conversation with patients. While it is possible that physicians and patients can discuss diet records, it is a cumbersome method that still fails to capture metabolic individuality. Thus, nutrition remains focused on global recommendations rather than treat-to-target individualized advice.

Watkins discussed the development of assays to quantify the blood levels of many important dietary nutrients, including fatty acids, sterols, amino acids, and other markers of nutrient status, including acylcarnitines and bile acids. These assays have been used to identify the relationship of blood measures of nutrients with metabolic outcomes including conversion to diabetes. The concept is that by measuring the blood levels of key nutrients directly, patients and physicians can have simpler, more accurate and productive discussions about nutrition therapy. Nutrition advice that is individualized and targeted, as opposed to global, can be given.

Watkins described results of this approach using baseline samples from the Insulin Resistance Atherosclerosis Study,[18] where they determined the association of baseline nutrient levels with risk for diabetes within five years. The study found strong positive associations of blood saturated fats and cholesterol synthesis intermediates, and strong negative associations between plant based fatty acids and sterols with diabetes risk. These observations are consistent with the nutrition guidance for diabetes prevention based on major clinical prevention trials.[19] Additionally, weak positive associations

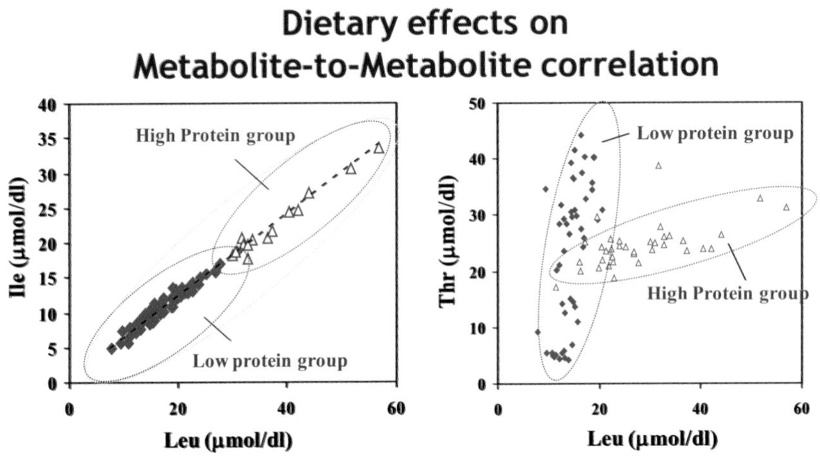

Dietary effects on Metabolite-to-Metabolite correlation

Figure 6. Opportunities to use disruptions in normal correlations among circulating free amino acids as biomarkers of health conditions, for example, effects of changes in dietary protein intake in the rat.

between branched chain amino acids and odd-chain acylcarnitines and diabetes risk were observed.

These results suggest that using a measurement strategy, as opposed to diet recall, could enable a personalized approach to nutrition, where individual needs can be measured and nutrient advice can be dispensed in a treat-to-target paradigm instead of as global recommendations.

Amino acid–based biomarkers for indicating nutritional and disease states

Takeshi Kimura (Ajinomoto Co., Inc., Japan) presented his group's innovative use of plasma amino acid profiles to screen for cancer and moderate malnutrition. He pointed out that, traditionally, biomarkers have typically been single molecules, the behaviors of which were pertinent to a phenotype or physiological state of interest. Modern metabolomic technologies, however, have made it possible to determine multiple metabolites and, thus, to facilitate capturing the status of specific biochemical pathways, offering unprecedented potential for generating biomarkers for various physiological states. Biomarkers generated in this manner have become important diagnostic criteria in various clinical areas; however, to date only a very small part of the information contained in the human metabolome has been used in the human health field.

Correlations of amino acid concentrations in various tissues have demonstrated that plasma amino acid patterns can serve as biomarkers for a number of diseases and physiological states (Fig. 6). The

generation of such metabolomics data, however, requires that procedures for collecting, handling, processing, and analyzing samples be validated and standardized.[20, 21]

Kimura noted that, after much testing, his group has developed a technology package to generate plasma amino acid–based markers for use in clinical settings. Their approach allows a single measurement of plasma amino acids to provide data on multiple biomarkers, each biomarker being a different amino acid profile. Both discriminative and surrogate markers can be generated depending on the target phenotype. Studies in rodent models indicate that this approach can also be used to distinguish protein malnutrition. Their approach has been used to generate biomarkers of risk for cancers of the stomach, lung, colon/rectum, prostate, and breast.[22] Research is ongoing to determine whether specific plasma amino acid profiles can be used as biomarkers for hepatitis, irritable bowel syndrome, and metabolic syndrome.[23, 24]

Kimura concluded by noting that, in April 2011, their technology was offered as a service to hospitals and clinics. As of July 2012, some 300 hospitals and clinics had adopted it as a blood test offered as an option to patients and healthy individuals.

Biomarkers of fruit and vegetable intake

Michelle C. McKinley (Queen's University, Belfast) reviewed ongoing work to identify blood-based biomarkers of fruit and vegetable consumption. Much of the evidence relating food and nutrient

Table 1. Results of clinical intervention trials with fruits and vegetables, showing the value of biomarkers as indicators of compliance[26]

Biomarker	Whole-diet studies ($n = 11$)		Mixed fruit/vegetable studies using counseling ($n = 16$)		Mixed fruit/vegetable studies using food provision ($n = 36$)	
	Measured biomaker	Reported increase (% use biomarker)	Measured biomaker	Reported increase (% use biomarker)	Measured biomaker	Reported increase (% use biomarker)
α-Carotene	7	3 (43)	14	12 (86)	21	16 (76)
β-Carotene	7	4 (57)	15	13 (87)	24	19 (79)
Lycopene	7	2 (29)	12	4 (33)	20	8 (40)
β-Cryptoxanthin	7	1 (14)	12	8 (67)	20	12 (60)
Lutein	4	3 (75)	7	5 (71)	16	11 (69)
Zeaxanthin	2	1 (50)	5	1 (20)	11	3 (27)
Lutein/zeaxanthin	3	2 (67)	5	3 (60)	5	4 (80)
Total carotenoids	5	2 (40)	6	4 (67)	2	1 (50)
Vitamin C	1	1 (100)	10	6 (60)	18	14 (78)
Urinary potassium	1	1 (100	2	0	2	1 (50)
Flavonoids	–	–	1	0	8	6 (75)

intake to chronic disease risk relies on information gathered by various dietary assessment techniques. Such techniques are prone to random and systematic measurement errors, which may attenuate observed diet–disease associations.[25] Therefore, biochemical biomarkers of dietary exposure are of great interest, as their use may improve the ranking of subjects for exposure to a particular food group or nutrient. Nutritional biomarkers also offer the possibility of an objective indicator of compliance with a particular dietary regimen in randomized controlled trials investigating the health effects of dietary modifications.

A systematic review of biomarkers of fruit and vegetable intake in human intervention studies[26] included over 90 intervention trials. This included studies of three types: (1) whole-diet intervention studies (advice to increase fruit and vegetable intake was one component of a whole diet approach); (2) mixed fruit and vegetable studies (interventions involving administration of more than one type of fruit or vegetable); and (3) individual fruit and vegetable intervention studies (involving increased consumption of a certain type of fruit or vegetable).

These studies showed that a panel of biomarkers (α- and β-carotene, vitamin C, lutein, zeaxanthin, and β-cryptoxanthin) had value as indicators of compliance in fruit and vegetable intervention trials (Table 1). With the possible exception of fruit-only intervention studies, where assessment of vitamin C status alone may suffice, it seems rarely possible to rely on assessment of a single biomarker as an indicator of change in fruit and vegetable intake. The review also pointed to the need for more dose–response data to elucidate the natures of the dose–response relationships of specific biomarkers.

McKinley noted that vitamin C and carotenoids were the most commonly measured biomarkers and that relatively few trials with mixed fruit and vegetables have used other biomarkers of fruit/vegetable consumption, for example, individual/total flavonoid status. She pointed to ongoing work involving novel use of panels of biomarkers, including plasma vitamin C, carotenoids, and possibly flavonoids to develop algorithms predictive of fruit and vegetable intake.[27] This approach seeks to further explore the compositional complexities of fruit and vegetables. She concluded that a more global consideration of a panel of potential biomarkers is likely to be more useful than single compounds.

Acknowledgments

The conference "Biomarkers in Nutrition: New Frontiers in Research and Application" was supported by the Sackler Institute for Nutrition Science.

Conflicts of interest

Takeshi Kimura is a corporate executive officer of, and owns stock in, Ajinomoto Co., Inc. The other authors declare no conflicts of interest.

References

1. Black, R.E., L.H. Allen, Z.A. Bhutta, *et al.* 2008. Maternal and child undernutrition: global and regional exposures and health consequences. *Lancet* **371:** 243–260.

2. Ramakrishnan, U. 2002. Prevalence of micronutrient malnutrition worldwide. *Nutr. Rev.* **60:** S46–S52.

3. Garcia, O.P., K.Z. Long & J.L. Rosado. 2009. Impact of micronutrient deficiencies on obesity. *Nutr. Rev.* **67:** 559–672.

4. Global Strategy on Diet, Physical Activity and Health. Geneva: World Health Organization. http://www.who.int/dietphysicalactivity/strategy/eb11344/strategy_english_web.pdf

5. Jehn, M & A. Brewis. 2009. Paradoxical malnutrition in mother-child pairs: untangling the phenomenon of over- and under-nutrition in underdeveloped economies. *Econ. Hum. Biol.* **7:** 28–35.

6. Hooper, L., K. Ashton, L.J. Harvey, *et al.* 2009. Assessing potential biomarkers of micronutrient status by using a systematic review methodology: methods. *Am. J. Clin. Nutr.* **89:** 1953S–1959S.

7. Raiten, D.J., S. Namasté, B. Brabin, *et al.* 2011. Executive summary–Biomarkers of Nutrition for Development: Building a Consensus. *Am. J. Clin. Nutr.* **94:** 633S–650S.

8. Institute of Medicine. 2010. *Evaluation of Biomarkers and Surrogate Endpoints in Chronic Disease.* Washington D.C. National Academies Press.

9. Combs, Jr., G.F. 2011. Determinants of selenium status in healthy adults. *Nutr. J.* **10:** 75–82.

10. Combs, Jr., G.F., *et al.* 2011. Differential Responses to selenomethionine supplementation by sex and genotype in healthy adults. *Br. J. Nutr.* Sep 22: 1–12.

11. Clark, L. *et al.* 1996. The Nutritional Prevention of Cancer with Selenium 1983–1993: a Randomized Clinical Trial. *J. Am. Med. Assoc.* **276:** 1957–1963.

12. Klein, E.A. 2011. Vitamin E and the risk of prostate cancer: the Selenium and Vitamin Cancer Prevention Trial (SELECT). *J. Am. Med. Asoc.* **306:** 1549–1556.

13. Bleys, J., A. Navas-Acien & E. Guallar. 2007. Serum selenium and diabetes in U.S. adults. *Diabetes Care* **30:** 829–834.

14. Stranges, S. *et al.* 2005. Effects of selenium supplementation on cardiovascular disease incidence and mortality: secondary analyses in a randomized clinical trial. *Am. J. Epidemiol.* **163:** 694–699.

15. Walston, J.D. 2012. Sarcopenia in older adults. *Curr. Opin. Rheumatol.* **24:** 623–627.

16. Studenski, S. 2012. Conference Proceedings: Evidence-Based Criteria for Sarcopenia with Clinically Important Weakness. *Semin Arthritis Rheum.* Sep 14. pii: S0049-0172(12)00180-1. doi: 10.1016/j.semarthrit.2012.07.007.

17. Van Kan, G.A. *et al.* 2011. Sarcopenia: biomarkers and imaging (International Conference on Sarcopenia research). *J. Nutr. Health Aging* **15:** 834–846.

18. Wagenknecht, L.E. *et al.* 1995. The insulin resistance atherosclerosis study (IRAS) objectives, design, and recruitment results. *Ann. Epidemiol.* **5:** 464–472.

19. American Diabetes Association. 2010. Standards of medical care in diabetes. *Diabetes Care* **33:** S11–S61.

20. Noguchi, Y. *et al.* 2006. Network analysis of plasma and tissue amino acids and the generation of an amino index for potential diagnostic use. *Am. J. Clin. Nutr.* **83:** 513S–519S.

21. Imaizumi, A. *et al.* 2012. Clinical Implementation of Metabolomics. In Metabolomics. U. Roessner, Ed. ISBN: 978-853-51-0046-1.

22. Miyagi, Y. *et al.* 2011. Plasma free amino acid profiling of five types of cancer patients and its application for early detection. *PLoS ONE* **6:** e24143.

23. Hisamatsu, T. *et al.* 2012. Novel, objective, multivariate biomarkers composed of plasma amino acid profiles for the diagnosis and assessment of inflammatory bowel disease. *PLoS ONE* **7:** e31131.

24. Yamakado, M. *et al.* 2012. Plasma amino acid profile is associated with visceral fat accumulation in obese Japanese subjects. *Clin. Obesity* **2:** 29–40.

25. Jenab, M. *et al.* 2009. Biomarkers in nutritional epidemiology: applications, needs and new horizons. *Hum. Genet.* **125:** 507–525.

26. Baldrick, F.R. *et al.* 2011. Biomarkers of fruit and vegetable intake in human intervention studies: a systematic review. *Crit. Rev. Food. Sci. Nutr.* **51:** 795–815.

27. Medical Research Council Research Portfolio: http://www.mrc.ac.uk/ResearchPortfolio/Grant/Record.htm?GrantRef=G0901793&CaseId=16175

Ann. N.Y. Acad. Sci. ISSN 0077-8923

ANNALS OF THE NEW YORK ACADEMY OF SCIENCES
Issue: Annals *Meeting Reports*

The new revolution in toxicology: The good, the bad, and the ugly

Myrtle Davis,[1] Kim Boekelheide,[2] Darrell R. Boverhof,[3] Gary Eichenbaum,[4] Thomas Hartung,[5] Michael P. Holsapple,[6] Thomas W. Jones,[7] Ann M. Richard,[8] and Paul B. Watkins[9]

[1]Toxicology and Pharmacology Branch, Developmental Therapeutics Program Division of Cancer Treatment and Diagnosis, The National Cancer Institute, National Institutes of Health, Bethesda, Maryland. [2]Deparment of Pathology and Laboratory Medicine, Brown University, Providence, Rhode Island. [3]Toxicology and Environmental Research and Consulting, The Dow Chemical Company, Midland, Michigan. [4]Department of Drug Safety Science, Johnson & Johnson Pharmaceutical R&D, LLC, Raritan, NJ. [5]Department of Environmental Health Sciences. Johns Hopkins Bloomberg School of Public Health, Baltimore, Maryland. [6]Battelle Memorial Institute, Columbus, Ohio. [7]Department of Toxicology and Pathology, Elil Lilly and Company, Indianapolis, Indiana. [8]National Center for Computational Toxicology, Environmental Protection Agency, Research Triangle Park, North Carolina. [9]Institute for Drug Safety Sciences, Hamner University of North Carolina, Research Triangle Park, North Carolina

Address for correspondence: Myrtle Davis, D.V.M., Ph.D., Toxicology and Pharmacology Branch, Developmental Therapeutics Program, Division of Cancer Treatment and Diagnosis, The National Cancer Institute, NIH Bethesda, MD 20852. davismillinm@mail.nih.gov

In 2007, the United States National Academy of Sciences issued a report entitled *Toxicity Testing in the 21st Century: A Vision and a Strategy.* The report reviewed the state of the science and outlined a strategy for the future of toxicity testing. One of the more significant components of the vision established by the report was an emphasis on toxicity testing in human rather than animal systems. In the context of drug development, it is critical that the tools used to accomplish this strategy are maximally capable of evaluating human risk. Since 2007, many advances toward implementation of this vision have been achieved, particularly with regard to safety assessment of new chemical entities intended for pharmaceutical use.

Keywords: toxicology; pharmaceuticals; testing

Introduction

Protection of human safety is a primary objective of toxicology research and risk management. In June 2007, the U.S. National Academy of Sciences released a report *Toxicity Testing in the 21st Century: A Vision and a Strategy.*[1] This report (Tox21C) included four main components: (1) chemical characterization, (2) toxicity pathways and targeted testing, (3) dose response and extrapolation modeling, and (4) human exposure data. The report outlined a new vision and strategy for toxicity testing that would be based primarily on human rather than general animal biology and would require substantially fewer or virtually no animals. There are clear promises and challenges associated with the vision, including the recognition that components of this vision are natural extensions of the evolution of toxicology science, and the challenging and long-standing debate

associated with a total reliance on cell-based systems and *in vitro* methods. While there are many topics and issues of interest to toxicologists, there are only a few that have the potential to have as great an impact on the science of toxicology as this vision. As such, it can be debated as to whether the report was the initial skirmish in what is now being called the *new revolution in toxicology*.

The realization of this vision will depend upon defining a series of toxicity pathways (e.g., cytotoxicity, cell proliferation, apoptosis, etc.) that can be monitored using medium- to high-throughput *in vitro* test systems—preferably based on human cells, cell lines, or tissues—and that are expected to provide a sufficiently comprehensive characterization of human risk and to reduce or eliminate the use of the apical endpoints currently collected through *in vivo* animal testing. It is acknowledged that an extraordinary amount of effort will be needed to

doi: 10.1111/nyas.12086

(1) determine the most informative set of toxicity pathways; (2) develop, validate, and implement the appropriate test systems; (3) create the necessary data management and computational tools; and (4) define how regulatory decision making will be adjusted to utilize these new data. While the report, sponsored by the U.S. Environmental Protection Agency, primarily focuses on the challenges of identifying, assessing, and managing the risks associated with human exposure to chemical agents found in the environment, there is passing reference to the potential of applying this revolutionary approach to toxicity testing in other applications, including pharmaceutical research and development. It is important to note that the report stops short of recommending expanding the testing requirements for pharmaceuticals, but interest in extending the Tox21C testing principles in that direction has clearly grown over the last several years. However, there has been very limited discussion regarding the inherent differences between how toxicity testing is applied to enable human pharmaceutical development and how it is used to support environmental decision making.

In October 2011, the New York Academy of Sciences hosted a conference entitled "The New Revolution in Toxicology: The Good, Bad and Ugly," sponsored by the Academy and Agilent Technologies, Cephalon, and Cyprotex, and promoted by the American College of Toxicology and the Society of Toxicology. This one-day conference attracted approximately 200 attendees, including experienced and new investigators; clinicians; toxicologists; and policy makers with multidisciplinary expertise from the fields of pharmacology, genetic and molecular toxicology, animal study design, drug discovery and development, computational chemistry, environmental law, cell and molecular biology, and pathology. The conference was particularly timely and exciting given the recent explosion of broadly applicable new models and technologies for assessing drug efficacy and toxicity. A primary goal of this symposium was to advance the discussion by focusing on how these differences might affect choices of appropriate test systems and how those systems would be applied in practice.

Implementation of the vision: beyond 2007

The symposium began with a session intended to capture reflections on the status of implementation of the 2007 NAS Report. The initial discussion began with a talk from Daniel Krewski (University of Ottawa). Krewski discussed data and provided rationale to support a broader, population-based approach to risk assessment. Relevant to this topic is a recent report from the Institute of Medicine (IOM) "Breast Cancer and the Environment: A Life Course Approach."[2] In this report, among the environmental factors reviewed, those most clearly associated with increased breast cancer risk in epidemiological studies were use of combination hormone therapy products, current use of oral contraceptives, exposure to ionizing radiation, overweight and obesity among postmenopausal women, and alcohol consumption. Krewski stressed that interactions between these factors must be integrated into any risk assessment strategy and tailored to inform the strategy. Along these lines, Christopher I. Li and colleagues (Fred Hutchinson Cancer Research Center) conducted an observational study of a subset of patients in the Women's Health Initiative (WHI) study, between 1993 and 1998, which included 87,724 postmenopausal women aged 50–79 years.[3] They reported that alcohol use is more strongly related to the risk of lobular carcinoma than to ductal carcinoma, and more strongly related to hormone receptor–positive breast cancer than to hormone receptor–negative breast cancer. Their results supported the previously identified association of alcohol consumption with hormone-positive breast cancer risk, as well as three previous case-control studies that identified a stronger association of alcohol with lobular carcinoma.

Krewski also discussed the use of a new approach to dose-response analysis that uses what is termed a *signal to noise crossover dose* (SNCD). The SNCD is defined as the dose where the additional risk is equal to the background noise (the difference between the upper and lower bounds of the two-sided 90% confidence interval on absolute risk) or a specified fraction thereof. In his published study, the National Toxicology Program (NTP) database was used as the basis for these analyses, which were performed using the Hill model. The analysis defined the SNCD as a promising approach that warrants further development for human health risk assessment.[4] Finally, Krewski concluded with a vision for the incorporation of systems biology that was offered a number of years ago in a publication authored by Stephen

W. Edwards and R. Julian Preston.[5] As described in the publication, it was proposed that systems approaches may provide a method for generating the type of quantitative mechanism of action data required for risk assessment.

Thomas Hartung (Johns Hopkins Bloomberg School of Public Health) discussed in his talk "Implementation of the NAS Vision" a compelling opinion on this topic. Hartung reflected on the perceived atmosphere created by the Tox21C report. Overall his opinion was that the report, indirectly suggests moving away from traditional (animal) testing toward modern technologies based on pathways of toxicity. The concept presented was rather simple in his view: there might be only a couple of hundred ways to harm a cell and these pathways of toxicity could be modeled in relatively simple cell tests, which can be run by robots. The goal is to develop a public database for such pathways, the human toxome, to enable scientific collaboration and exchange. Hartung also mentioned that there is a continuously growing awareness—not necessarily always described as excitement—about Tox21C in stakeholder groups. It was first embraced by scientists in the United States—it was difficult to find a U.S. toxicology meeting over the last few years that did not have it as a topic. The U.S. Society of Toxicology instantly started a series of comments in their journal. Most importantly, the U.S. agencies followed fast on the 2007 NAS/NRC report: the Tox21C alliance in 2008 (a paper in *Science* first-authored by Francis Collins[6]); the EPA made it their chemical testing paradigm in 2009;[7] the Food and Drug Administration (FDA) followed most evidently with the *Science* article by Margret Hamburg in 2011.[8] The chemical and consumer product industry got engaged, e.g., the Human Toxicology Project Consortium,[9] as did the pharmaceutical industry, somewhat more reluctantly (in his opinion). In Europe, a reaction to the paradigm has been delayed, with some adaptation of the vocabulary but not necessarily an embrace of the new approach. He did not view the current strategies as alternative methods under a new name. However, Hartung felt that interest is strongly increasing in Europe.

Hartung was also clear that Tox21C suggests more than just movement toward databases of pathways of toxicity. One big problem is that the respective science is still emerging. The call for mechanism-based approaches has been around for a while. The

new concept that may be articulated going forward is a change in resolution to molecularly defined pathways. The new technologies (especially omics) may allow this. He followed by stating that what is needed is the human toxome, viewable as a comprehensive pathway list, annotation of cell types, references to species differences (or source of elucidation), toxicant classes and hazards to these pathways, an integration of information in systems toxicology approaches, the *in vitro–in vivo* extrapolation by reversed dosimetry, and finally, a means to make sense of the data, most likely in a probabilistic way. Hartung presented a list of the most notable activities:

- The EPA launched its ToxCast program based on available high-throughput tests; the EPA made Tox21C their official toxicity testing paradigm for chemicals in 2009.
- The Tox21C alliance of the EPA, the National Insitute of Environmental Health Sciences (NIEHS), the National Cartography and Geospatial Center (NCGC), and the FDA extended this to more chemicals.
- Case study approaches at the Hamner Institute, which was originally created by the chemical industry.
- The Human Toxicology Project Consortium (seven global companies and three stakeholders including the Center for Alternatives to Animal Testing (CAAT)).
- The Human Toxome Project led by CAAT and financed by an NIH Transformative Research Award (this very competitive and prestigious grant is given for projects that have the potential to drive change). The project involves ToxCast, the Hamner Institute, Agilent, and several members of the Tox21C panel.
- The Organisation for Economic Co-operation and Development (OECD) has embraced this in their new *adverse outcome pathway* concept.

Hartung discussed what he considered to be the most advanced regulatory application: a proposal to use some of the high-throughput assays from ToxCast to prioritize endocrine disruptor screening (EDSP21 program) as a representative example of this advanced use. However, he emphasized that this proposal is a work in progress and has already generated some resistance. He also mentioned that OECD has embraced some of the concepts under the label of adverse outcome pathways. Lastly, Hartung surmised that, early on, there is clearly a need for a process to qualify the new approaches as a critical component of their development and implementation. Formal validation as developed for the first generation of alternative methods can only partially serve this purpose. For this reason, the Evidence-based Toxicology Collaboration (EBTC) was created in the United States and Europe in 2011 and 2012, respectively (www.ebtox.com). This collaboration of representatives from agencies, industry, academia, and stakeholder groups aims to apply the tools developed by evidence-based medicine to toxicology. The EBTC secretariat is run by CAAT, and the first conference was held in early 2012 and hosted by the U.S. EPA. Working groups have started to address pertinent issues and methodologies. Taken together, Tox21C and its implementation activities including the human toxome and the EBTC promise a credible approach to revamp regulatory toxicology.

The vision for toxicity testing in the 21st century

Michael P. Holsapple's (ATS, Battelle Memorial Institute) talk, "Vision for Toxicity Testing in the 21st Century (Tox21C): Promises, Challenges, and Progress," began with a few excerpts from the report that he thought represented the stimulus for the Tox21C vision and strategy: " . . . transformative paradigm shift and . . . new methods in computational biology and a comprehensive array of *in vitro* tests based on human biology." He echoed some of Hartung's comments about the importance of collaborative activities, such as the EBTC (e.g., Holsapple noted that he serves as a member of the steering team) and the Human Toxicology Project Consortium (e.g., he indicated that he was a co-author of the 2010 HTPC workshop report, which is in press).[10] Holsapple indicated that he would focus his remaining time describing two other activities with which he has had some personal experience, and which could serve as additional perspectives on approaches to advance the Tox21C vision and strategy. He then presented a brief overview of the ILSI Health and Environmental Sciences Institute (HESI) Risk Assessment for the 21st Century (RISK21) program. Holsapple indicated that the impetus for RISK21 included the NAS report "Science and Decision: Advancing Risk Assessment"[11] and the 2007 NAS Tox21C report, and that the stimulus for RISK21

included the ever increasing development and use of new technologies that would impact risk assessment, such as the following: (1) high-content technologies (e.g., genomics, proteomics, metabonomics); (2) high density approaches such as high throughput toxicity assays; (3) sensitive new analytical chemistry techniques; and (4) increasingly detailed knowledge of cellular exposure (PBPK modeling methods).

The leaders of the HESI RISK21 program recognize that there has been a lack of consensus on how best to use and incorporate the information from these new methods into quantitative risk assessments, and that there was an opportunity to provide broad scientific leadership to develop credible approaches and to suggest changes in policies. The vision for RISK21 was to initiate and stimulate a proactive and constructive dialogue among experts from industry, academia, government, and other stakeholders to identify the key advancements in 21st century risk assessment. The RISK21 program was structured around four working groups: (1) dose-response—establish a unified approach to dose-response assessment that builds on the existing mode of action and key events dose response framework (KEDRF) to quantitatively incorporate dose-response information, and to address technical issues regarding *in vitro* to *in vivo* extrapolation; (2) cumulative risk—define and develop critical elements of a transparent, consistent, pragmatic, scientific approach for assessing health risks of combined exposures to multiple chemicals in the context of other stressors; (3) integrated evaluation strategies—establish a process that provides flexible, rapid, efficient, transparent, and cost-effective scientifically sound approaches for creating and interpreting data relevant to decision making that protects health; and (4) exposure science—propose approaches for using new technologies to improve characterization of real-world exposures and provide the data-driven evidence base for 21st century exposure modeling and risk assessment.

As a segue into his final topic, Holsapple again referred to the previously described paper in *Science* by FDA commissioner Margaret Hamburg, where she noted, "We need better predictive models to identify concerns earlier in the (drug) development process to reduce time and costs. We also need to modernize the tools used to assess emerging concerns about potential risks from food and other exposures", and

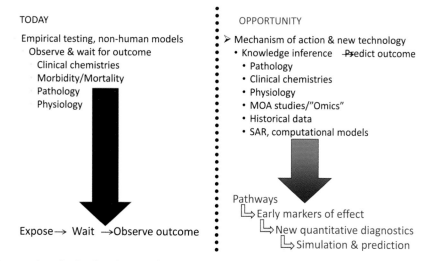

Figure 1. Menu- and mechanism-based approaches in toxicity testing. Transitioning to mechanism-based approaches could improve predictivity and insight, combining an initial triage step, using multi-scale imaging and modeling, to quantify dosimetry and identify highest priority concerns, with subsequent targeted steps using integrated *in vitro* and *in vivo* methods to assess priority concerns.

to the 2011 FDA Strategic Plan, which listed the top priority as "Modernize toxicology to enhance product safety." Holsapple then described the Battelle Multi-Scale Toxicology Initiative (MSTI; Fig. 1). He noted that the vision for MSTI was based on the belief that mechanism-based approaches could speed testing and improve accuracy in picking winners. The stated goal of the MSTI concept is to improve predictivity, insight, and translational value via skillful integration of data at all scales. The concept was to move away from what he described as current menu-driven approaches into an approach that has (1) an initial triage step to quantify dosimetry and identify highest priority concerns using multi-scale imaging and modeling; and (2) subsequent, targeted steps to interrogate and assess priority concerns using integrated *in vivo* and *in vitro* methods. Holsapple concluded by emphasizing that toxicology, especially in the context of toxicity testing during the drug development process, needs to be updated, and by noting that the NAS Tox21C vision and strategy has the promise to address that need. However, he cautioned that many challenges remain to realize that promise, and that in spite of government buy-in from the EPA, NIH, NIEHS/NTP, and FDA, the need for 21st century validation tools and proof-of-concept projects—especially for risk assessment and regulatory science—must be recognized.

Legal acceptance of the Tox21C approach

E. Donald Elliott, J.D., rounded out the session by presenting an intriguing review of some of the legal consequences of the changes in the approach to safety assessment in his presentation "Mapping the Pathway to Legal Acceptance of Pathway-based Toxicological Data: How to Climb Mt. Krewski". Elliott started out providing a description of our legal system and some important realizations. Our system, as he described, is one based on judicial precedent set by generalist judges, and tends to be conservative in allowing new scientific paradigms and techniques to be accepted in court. There is no single legal standard that pathway-based (non-apical) data must satisfy; legal hurdles differ in different contexts. Standards for legal acceptance of emerging scientific information by courts of general jurisdiction with lay juries are high, whereas administrative agencies have more discretion. Judges act as gatekeepers to keep new scientific information from reaching juries until it has been shown to be reliable and applicable (the Daubert rule, named for the Supreme Court case that created it). Elliott went on to contrast the judicial system with administrative agencies. He made the important distinction that administrative agencies are considered experts and that they may consider emerging scientific information along with other information provided that their final decisions are supported by "substantial evidence on the record as a whole." Non-expert judges in reviewing courts do not rule on the admissibility of individual pieces of evidence before administrative agencies. Therefore, agencies have broad discretion to consider pathway-based toxicological information along with other evidence in making weight of evidence determinations, and have already begun to do so. Elliott provided the approval of chemical dispersants for use against the Deep Water Horizon oil spill as a recent example in which the EPA relied in part on pathway-based testing. In general, the legal system is more lenient in accepting new information where the costs of a false positive or false negative are relatively low. He left the audience with an important conclusion to ponder: broader acceptance of pathway-based toxicology in contexts where the stakes are higher (such a approving a new drug or banning a pesticide) will depend on developing side-by-side comparisons to the predictive value of traditional data.

Key differences and challenges between safety and risk assessment

The intent of Session II was to discuss the differences and challenges between safety and risk assessment, and the use of *in vitro* Assays. The session began with an insightful talk entitled "Predictive Toxicology at Abbott in Early Discovery: A Critical Review of Successes and Failures over an 8-Year Period" provided by Eric Blomme (Abbott Laboratories). Blomme reviewed experience at Abbott over the last eight years using examples to illustrate the strengths, weaknesses, limitations, and optimal application of these technologies during lead optimization and candidate selection. Finally Blomme discussed several recent promising additions to the toxicologist's toolbox and some key challenges facing toxicologists in the pharmaceutical industry in the near future.

Thomas W. Jones (Eli Lilly and Company) provided an inspiring contrast of ideals. Jones started by characterizing Tox21C as a response rather than an answer and pointed out the fact that the committee itself describes the report as a "long-range vision and strategic plan". He pointed out the salient features of the report and focused on other important factors that frame the discussion. He reiterated that the expectation that the products of the revolutions in biology and biotechnology can be used to transform toxicity testing from a system based on whole animal studies to one founded primarily on *in vitro* methods using cells of human origin is in fact a hypothesis to be tested and presents an open question regarding applications in other areas of toxicology supporting human risk assessment (Fig. 3). Jones made several points about how we might best apply a human cell-based nonclinical testing scheme, similar to the one envisioned in the Tox21C report, in human pharmaceutical discovery and development to inform and improve decision making. He argued that it is critical to be sure that we have a common understanding of how the current testing paradigm is applied and how it performs, and made these key distinctions: (1) if risk is realized early in a drug development program, then less clinical data will be available providing context and balance to the decision makers regarding benefit; (2) there needs to be a balance in the revelation of risk and benefit; (3) the science being done to realize the potential of Tox21C could benefit the pharmaceutical industry—however, the experience of the pharmaceutical industry, where human clinical testing is a routine part of the development process, will inform how to apply the approaches outlined in Tox21C to other settings where human risk-based decisions are being made; and (4) the tools needed to support the safe testing of human pharmaceuticals are likely to be different from those needed for environmental decision making.

Jones addressed several popular perceptions and myths. One perception he discussed was that the pharmaceutical industry needs better ways to recognize safety risks in order to make decisions earlier. Along these lines, he mentioned the current systematic use of animal models as a sentinel measure of safety risk. In Figure 2, Jones outlined the utility of the current animal-based approach and concordance. For example, the sensitivity of fecal occult blood for detection of colon cancer is about 66% and the sensitivity of prostate-specific antigen (PSA) screening for detection of prostate cancer is about 72%. Although animal to human translation varies by toxicity and target tissue, pre–first-in-human toxicology studies reveal nearly all toxicities that are relevant to humans.

Tox21 and ToxCast chemical landscapes

Ann M. Richard (U.S. Environmental Protection Agency) reviewed the U.S. EPA's ToxCast project and the related multi-agency Tox21C projects. These

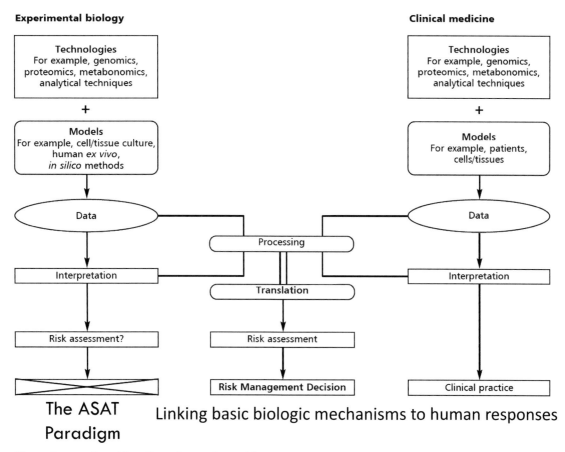

Figure 2. An outline of the utility and concordance of the current systematic use of animal models as sentinel measures of safety risk. Pre–first-in-human toxicology studies reveal nearly all toxicities that are relevant to humans, although translation varies by target tissue and toxicity.

projects are employing high-throughput technologies to screen hundreds to thousands of chemicals in hundreds of assays, probing a wide diversity of biological targets, pathways and mechanisms for use in predicting *in vivo* toxicity. The ToxCast chemical library consists of 960 unique chemicals (including Phase I and II), with 100 recently added to this total, and was constructed to span a diverse range of chemical structures and use categories. This library is fully incorporated into the EPA's approximately 4000 chemicals contributing to the larger, more diverse Tox21 chemical library (totaling 10,000). These chemical libraries represent central pillars of the ToxCast and Tox21 projects and are unprecedented in their scope, structural diversity, multiple use scenarios (pesticides, industrial, food-use, drugs, etc.), and chemical feature characteristics in relation to toxicology. Chemical databases

built to support these efforts consist of high quality DSSTox chemical structures and generic substance descriptions linked to curated test sample information (supplier, lot, batch, water content, analytical QC). Cheminformatics, feature and property profiling, and a priori and interactive categorization of these libraries in relation to biological activity will serve as essential components of toxicity prediction strategies. Finally, Richard mentioned the DSSTox project, which provides primary chemical library design and information management for the ToxCast and Tox21 projects.

The future of toxicology in drug development

David Jacobson-Kram (Food and Drug Administration) began with a reminder that toxicology testing serves several important needs in drug development. In the earliest stages, toxicology data

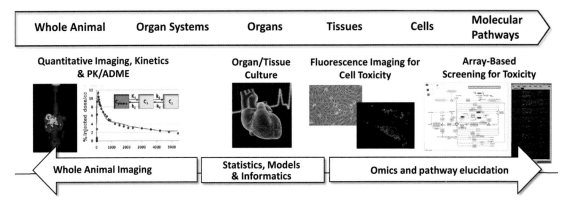

Figure 3. Different levels of toxicity analysis and current techniques for engaging each level. Despite biotechnical advances in many areas supporting human risk assessment, it remains to be definitively determined whether *in vitro* methods using cells of human origin can effectively supplant systems based on whole animal studies.

are used to determine the maximum recommended doses for first-in-man Phase I studies. This information can also help to identify potential toxicities and can also specify maximum stopping doses. Toxicology studies can also identify certain risks that cannot be studied in clinical trials: potential for genetic damage (genotoxicity), carcinogenicity, teratogenicity, and risks from long term exposures. *In vitro* studies have been useful in identifying hazards; for example, the Ames assay provides information on whether a drug candidate can induce gene mutations, and the hERG assay is useful in determining if a candidate pharmaceutical has potential for QT-interval prolongation. While hazard assessment is an important aspect in drug development, ultimately hazards must be linked to exposures in order to quantify risks. Use of *in vitro* assays in risk assessment is challenging primarily because of the difficulty in modeling absorption, distribution, metabolism, and excretion (ADME). Although not insurmountable, use of *in vitro* assays for all aspects of drug development will require new methods and better understanding of ADME processes.

Application of the strategy and development of the tools

Kim Boekelheide (Brown University) delivered an insightful review entitled "Toxicity Testing in the 21st Century: A Toxicologic Pathology Perspective Using Testis Toxicity as an Example." Boekelheide stated that the European Union has been a leader in advancing alternative testing strategies through legislative action and funding the development of new

testing paradigms. In 2004, a group from Unilever in the United Kingdom conceptualized a new testing approach entitled Assuring Safety without Animal Testing (ASAT). The ASAT initiative sought to tie how human disease processes are identified in the clinic together with the development of human relevant *in vitro* mechanistic data, bypassing the current need for toxicity testing in animals. The focus and outcome of an ASAT workshop held June 15–17, 2011, on "Recent Developments and Insights into the Role of Xenobiotics in Male Reproductive Toxicity" were reviewed by Boekelheide. The key questions for this initiative were clear: (1) Does xenobiotic exposure cause testicular toxicity in men? (2) If not well understood, what are the data gaps? (3) Can testicular toxicity be studied in animals and are the results sufficiently predictive for humans? and (4) Can testicular toxicity be studied *in vitro* with sufficient reliability to identify the relevant mechanisms of action for humans?

To address the first question, known examples of xenobiotic-induced testicular toxicity were reviewed along with the limitations of current diagnostic approaches. The diagnosis of testicular toxicity is currently measured using semen parameters and serum hormones, like inhibin B and FSH. These biomarkers have significant limitations, including a delay between exposure and biomarker alteration, a highly variable measure of effect, and reliance on epidemiological associations. He concluded that current approaches to understanding the etiology of human male toxicant-induced testicular injury are insensitive, highly variable, and lack diagnostic specificity. The need for new tools was identified as a data gap (question 2), and the potential for using modern molecular approaches, such as serum assessment of testis-specific microRNAs and monitoring sperm molecular biomarkers, including sperm mRNA transcripts and DNA methylation marks, may address this data gap.

The potential and limitations of animal experiments to inform regarding the human response (question 3) was highlighted using phthalate-induced effects in fetal testis as an example. Characteristics of fetal testicular responses to phthalates were compared between human, mouse, and rat. Fetal rat testis responds to phthalate exposure by suppressing steroidogenesis in the Leydig cells, resulting in lower fetal testosterone levels. On the other hand, recent studies have shown that both human and mouse fetal Leydig cells are resistant to these phthalate-induced anti-androgenic effects.[12,13] Therefore rats, but not humans or mice, would be expected to show effects of lowered fetal testicular testosterone production, including reduced steroidogenic gene expression, shortened anogenital distance, nipple/areola retention, hypospadias, cryptorchidism, and Leydig cell hyperplasia. Interestingly, all three species share phthalate-induced alterations in fetal seminiferous cords, including the induction of multinucleated germ cells. This example makes the point that animal models of interactions of xenobiotics with testicular function may be limited by uncertainties about their relevance to humans.[14]

Regarding the last question, *in vitro* approaches to evaluating testicular function have so far been limited by the inability to recapitulate spermatogenesis *in vitro*, a complex process unique to the testis. The ASAT workshop focused on revolutionizing testis toxicity testing approaches, including the main recommendation to design a functioning "testis in a petri dish" capable of spermatogenesis. Limitations to developing this novel approach were cited, including the lack of relevant human cell lines, difficulties in assessing the efficiency of spermatogenesis, difficulties in incorporating the needed paracrine interactions, and finally, a reliance on cell-specific apical endpoints rather than developing insight into toxicant-associated modes of action. The workshop proposed developing a human stem cell differentiation model of spermatogenesis with stem cells progressing through meiosis and giving rise to haploid spermatids, combined with bioengineering ap-

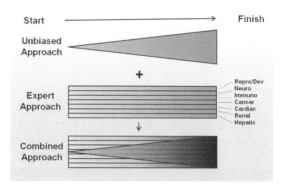

Figure 4. The development of new *in vitro* approaches to toxicity testing. The most thorough scope of toxicity response may be best evaluated through a balanced combination of unbiased testing and pathway-specific expert-driven testing.

proaches to build three-dimensional scaffolds for the assembly of the multiple cell types needed for successful spermatogenesis. Ultimately, the goal is to identify toxicants that interrupt this process using high-throughput *in vitro* tools.

The presentation ended with a broad view of approaches to development of the *in vitro* tests of the future. Major concerns about the validity of *in vitro* models were mentioned, including the requirement for key tissue-specific pathways, the ability of the *in vitro* system to recapitulate *in vivo* biology, and the complexity of interacting cell types. Boekelheide argued that the development of the new *in vitro* approaches might be best served with a balanced combination of unbiased testing and pathway-specific expert-driven testing, so that the most complete landscape of toxicity response could be evaluated (Fig. 4).

Evaluating injection site tolerability prior to testing in humans

Gary Eichenbaum (Janssen Research and Development) discussed the importance of evaluating and minimizing the infusion site irritation potential of intravenous formulations prior to testing new molecular entities in humans. Several promising advances that he discussed include the use of nonclinical *in silico* and *in vitro* models to screen candidate formulations and thereby enable a reduced use of *in vivo* nonclinical models. If an intravenous formulation causes severe injection site reactions, these adverse effects may pose significant challenges for the nonclinical safety assessment of systemic toxicity as well as the clinical development and use of the compound. The extent to which infusion site reactions may limit further development depends in part on the severity and root cause of the adverse effects, as well as the dose/concentration response and the margin of safety. The talk focused on *in vitro* and *in vivo* nonclinical strategies and models for developing, evaluating and optimizing intravenous formulations of compounds that have the potential to cause local toxicity following injection.

Eichenbaum reviewed the common types of injection site reactions (chemical and mechanical) and the key factors for chemically mediated injection site reactions: formulation concentration, solubility, and rate of injection. The common types listed include:

- Inflammation/irritation of the vessel at the injection site.
- Endothelial hyperplasia/intimal edema at injection site.
- Necrosis of the venous wall with mixed inflammatory cell infiltration/abscessation at the injection site.
- Thrombosis at the injection site.
- Necrosis with abscessation and adjacent inflammation at the entry point.
- Interstitial pneumonitis.
- Area of necrosis and inflammation in the lungs (probable infarct).
- Thromboembolus in the lungs.

He then reviewed targeted safety testing and discussed several nonclinical models to evaluate potential for infusion site reactions, such as a static kinetic solubility model; a dynamic solubility model; plasma solubility; *in vitro* hemolysis; cell based models; hen egg chorioalloantoic membrane model; and nonclinical *in vivo* infusion models.

Lastly, Eichenbaum provided a case example of a compound with limited solubility at neutral pH with potential to cause infusion site reactions at therapeutic concentrations. He presented a multifactorial approach to optimize and evaluate key parameters: test article concentration, pH, buffers, excipient types, excipient weight percentage, and dosing volume and rate. In this example, an *in silico* precipitation index model was developed to support infusion site safety assessment and the translation of nonclinical data to human situations. The model predicts that there may be some subtle differences

in species sensitivity (dog > rat > rabbit), but that the preclinical models should be fairly predictive despite the physiological differences.

A schema was presented for setting doses/concentrations that are more likely to be well tolerated in the nonclinical species and in humans. The *in silico* model results predict that if the plasma solubility is ≤ 0.05 mg/mL, then the formulation concentration should be ≤ 0.5 mg/mL to reduce the probability of an infusion site reaction. If the plasma solubility is between 0.05 and 0.1 mg/mL, then formulation concentrations up to 1.5 mg/mL are not likely to cause precipitation-mediated irritation. If the plasma solubility is >1 mg/mL, the model predicts minimal likelihood for precipitation or irritation up to concentrations of 10 mg/mL in the dosing solution. These predictions were in agreement with *in vivo* results with the example compound that was presented and other low solubility compounds that have been evaluated. The intravenous formulation that was selected from these assessments as part of the case-example was well tolerated in humans and achieved target exposures for efficacy.

The conclusions of the presentation were (1) an *in silico* and several *in vitro* models can help to screen and identify candidate formulations with reduced/limited infusion site irritation potential of low solubility compounds; (2) *in vivo* models provide additional and important information about risk for infusion site irritation of candidate formulations, but species differences must be considered; and (3) staged application of these models can support the successful optimization of intravenous formulations that have reduced risk for injection site irritation in humans, and at the same time reduce the number of *in vivo* nonclinical studies required.

Integrated cell signaling in toxicology

Many human diseases are a consequence of aberrant development of tissues in which the transition of adult stem cells into their appropriate differentiated cell type lineages and designated niches within a tissue has been interrupted by either genetic or epigenetic events. These molecular events often result in dysfunctional tissues or, as in the case of cancer, the development of tumors that bypass all normal cybernetic control mechanisms. These control mechanisms require homeostatic-regulated gene expression through highly coordinated networks of extracellular, intercellular, and intracellular signaling events within and between the cells of a tissue. Brad L. Upham (Michigan State University) proposed that gap junction intercellular channels are critical in modulating the levels of low molecular weight second messengers needed for the transduction of an external signal to the nucleus in the expression of genes essential to the normal maintenance of a tissue. Thus, any comprehensive systems biology approach to understanding the role of signaling in toxicology must also include gap junctions, as aberrant gap junctions have been clearly implicated in many human diseases.

Idiosyncratic hepatotoxicity: from human to mouse to computer

Paul B. Watkins, (The Hamner University of North Carolina Institute for Drug Safety Sciences (IDSS)) discussed drug induced liver injury (DILI), a major adverse drug event that leads to termination of clinical development programs and regulatory actions including failure to approve for marketing, restricted indications, and withdrawal from the marketplace. The type of DILI that is most problematic is idiosyncratic, meaning that only a very small fraction of treated patients are susceptible to the DILI. Current preclinical models, even humanized ones, do not reliably identify molecules that have this liability, and conversely, predict liabilities in molecules that are in fact quite safe for the liver. Reliable preclinical testing will probably not be developed until there is greater understanding of the mechanisms underlying idiosyncratic DILI. Based on the belief that the best models to study DILI are the people who have actually experienced it, there are two major efforts underway to create registries and tissue banks from these rare individuals: the Severe Adverse Events Consortium supported by industry and the Drug Induced Liver Injury Network supported by the National Institutes of Health. Genome-wide association analyses and whole exome/whole genome sequencing of certain DILI cases are well underway and appear promising. However, it has become clear that preclinical experimental approaches are also needed to both provide biological plausibility for associations observed and to generate specific hypotheses that can be tested with the genetic data. Ongoing approaches at the IDSS include chemoinformatic analysis of implicated drugs, use of panels of inbred and genetically

defined mice, and organotypic liver cultures including systems derived from induced pluripotent stem cells obtained from patients who have experienced DILI. The IDSS also leads DILIsim, a public–private partnership that involves scientists from 11 major pharmaceutical companies and the FDA, with the goal of integrating multiple streams of data into an *in silico* model that would explain and ultimately predict the hepatoxic potential of new drug candidates in humans.

Application of the strategy and development of the tools

Kyle Kolaja (Roche) proposed that one possible means to bridge the gap between late-stage assessments of safety-related organ toxicities and early discovery is through the use of human pluripotent stem cell–derived tissues, which afford improved cellular systems that replicate the critical functional aspects of intact tissues. The combination of stem cell–derived tissues with small-scale assays can ensure these models are amenable to high-throughput, low compound usage assays, and thus have utility in drug discovery toxicology. He presented published work that focused on stem cell–derived cardiomyocytes, characterizing these cells molecularly and functionally as a novel model of pro-arrhythmia prediction.

Darrell R. Boverhof (The Dow Chemical Company) discussed the promises and challenges of applying toxicogenomics and *in vitro* technologies for the assessment of chemical sensitization potential. He highlighted how advances in molecular and cellular biology are providing tools to modernize our approaches to chemical hazard assessment, including that for skin sensitization. Dr. Boverhof has been researching the application of toxicogenomics and *in vitro* assays for assessing the sensitization potential of chemicals. Determination of the skin sensitization potential of industrial chemicals, agrochemicals, and cosmetics is crucial for defining their safe handling and use. The mouse local lymph node assay (LLNA) has emerged as the preferred *in vivo* assay for this evaluation; however, the assay has certain limitations including the use of radioactivity, poor specificity with certain chemistries (false positives), and the inability to distinguish between different classes of sensitizers, namely skin and respiratory sensitizers. To address these limitations, researchers have been exploring the application of toxicogenomics to the LLNA to provide enhanced endpoints for the assessment of chemical sensitization potential. Data generated to date indicate that toxicogenomic responses are providing increased insight into the cellular and molecular mechanisms of skin sensitization, which may increase the specificity and extend the utility of the LLNA.[15–18] In parallel with these efforts, research is being conducted on the development and application of *in vitro* assays for predicting skin sensitization potential. Recent regulations (e.g., the EU Cosmetics Directive), as well as responsible stewardship, have pushed the development of non-animal approaches that can effectively predict skin sensitization potential for new chemical entities. These assays have built upon our current understanding of the molecular and cellular events involved in the acquisition of skin sensitization, and are showing promise for providing non-animal alternatives for characterization of skin sensitizing chemicals.[16]

Conclusions: Are we getting there?

The broad aims of this conference were (1) to provide a forum to discuss the recent advances in toxicity testing, their application to pharmaceutical discovery and development, and their relevance to safety assessment; (2) to bring together leading scientists from different disciplines to encourage interdisciplinary thinking; (3) to encourage outstanding junior scientists, students, and post-docs to pursue research in this promising field; (4) to provide networking opportunities among scientists and guests; and (5) to encourage collaborations to advance science. Conference speakers and audience participants achieved these aims through presentations and engaging discussion. At the end of the conference, participants were encouraged to critically explore questions such as:

- What was the initial response to the 2007 National Academy of Sciences report *Toxicity Testing in the 21st Century: A Vision and a Strategy*, and how did this vary across sectors?
- What has been achieved since the publication of the report?
- How has implementation of the recommendations in the report been achieved across sectors?
- What have been the strengths, weaknesses, limitations, and optimal applications of these technologies during lead optimization and candidate selection?
- What recent promising additions to the toxicologist's toolbox have emerged?
- What are key challenges facing toxicologists in the pharmaceutical industry in the near future?
- What lessons can we draw from the success of the report, and from the criticisms made against it?
- What recommendations can we adapt across the pharmaceutical industry?

Some of these questions have been the subject of follow-on initiatives and will be a constant source of ongoing discussions.

Conflicts of interest

The authors declare no conflicts of interest.

References

1. Committee on Toxicity Testing and Assessment of Environmental Agents, National Research Council. 2011 *Toxicity Testing in the 21st Century: A Vision and a Strategy*. National Academies Press. Washington D.C.
2. Committee on Breast Cancer and the Environment: The Scientific Evidence, Research Methodology, and Future Directions; Institute of Medicine. 2012. *Breast Cancer and the Environment: A Life Course Approach*. National Academies Press. Washington D.C.
3. Li, C.I. *et al.* 2010. Alcohol consumption and risk of postmenopausal breast cancer by subtype: the women's health initiative observational study. *J Natl Cancer Inst.* **102:** 1422–1431.
4. Sand, S., C.J. Portier & D. Krewski. 2011. A signal-to-noise crossover dose as the point of departure for health risk assessment. *Environ Health Perspect.* **119:** 1766–1774.
5. Edwards, S.W. & R.J. Preston. 2008. Systems biology and mode of action based risk assessment. *Toxicol Sci.* **106:** 312–318.
6. Collins, F.S., G.M. Gray & J.R. Bucher. 2008. Toxicology. Transforming environmental health protection. *Science* **319:** 906–907.
7. Firestone, M.R. *et al.* 2010. The U.S. Environmental Protection Agency Strategic Plan for Evaluating the Toxicity of Chemicals. *J. Toxicol. Environ. Health B. Cri. Rev.* **13:** 139–162.
8. Hamburg, M.A. 2011. Advancing Regulatory Science. *Science* **331:** 987–987.
9. Seidle, T. & M.L. Stephens. 2009. Bringing toxicology into the 21st century: A global call to action. *Toxicol in vitro.* **23:** 1576–1579.
10. Stephens, M.L. *et al.* 2012. Accelerating the development of 21st-century toxicology: outcome of a Human Toxicology Project Consortium workshop. *Toxicol. Sci.* **125:** 327–334.

11. Committee on Improving Risk Analysis Approaches Used by the U.S. EPA, National Research Council. 2009. *Science and Decisions: Advancing Risk Assessment.* National Academies Press. Washington D.C.

12. Mitchell, R.T. *et al.* 2012. Do phthalates affect steroidogenesis by the human fetal testis? Exposure of human fetal testis xenografts to di-n-butyl phthalate. *J Clin Endocrinol Metab.* **97:** E341–348.

13. Heger, N.E. *et al.* 2012. Human fetal testis xenografts are resistant to phthalate-induced endocrine disruption. *Environ Health Perspect.* **120:** 1137–1143.

14. Anson, B.D., K.L. Kolaja & T.J. Kamp. 2011. Opportunities for use of human iPS cells in predictive toxicology. *Clin Pharmacol Ther.* **89:** 754–758.

15. Boverhof, D.R. 2009. I. Evaluation of a toxicogenomic approach to the local lymph node assay (LLNA). *Toxicol Sci.* **107:** 427–539.

16. Ku, H.O. *et al.* 2011. Pathway analysis of gene expression in local lymph nodes draining skin exposed to three different sensitizers. *J Appl Toxicol.* **31:** 455–462.

17. Adenuga, D. *et al.* 2012. Differential gene expression responses distinguish contact and respiratory sensitizers and nonsensitizing irritants in the local lymph node assay. *Toxicol Sci.* **126:** 413–425.

18. Aeby P. *et al.* 2010. Identifying and characterizing chemical skin sensitizers without animal testing: Colipa's research and method development program. *Toxicol* In Vitro. **24:** 1465–1473.

Additional reading

Abassi, Y.A., B. Xi, *et al.* 2012. Dynamic monitoring of beating periodicity of stem cell-derived cardiomyocytes as a predictive tool for preclinical safety assessment. *Br. J. Pharmacol.* **165:** 1424–1441.

Bolt, H.M. & J.G. Hengstler. 2008. Most cited articles in the *Archives of Toxicology*: the debate about possibilities and limitations of *in vitro* toxicity tests and replacement of *in vivo* studies. *Arch. Toxicol.* **82:** 881–883.

Brendler-Schwaab, S.Y., P. Schmezer, *et al.* 1994. Cells of different tissues for *in vitro* and *in vivo* studies in toxicology: Compilation of isolation methods. *Toxicol.* In Vitro **8:** 1285–1302.

Charlton, J.A. & N.L. Simmons. 1993. Established human renal cell lines: Phenotypic characteristics define suitability for use in *in vitro* models for predictive toxicology. *Toxicol.* In Vitro **7:** 129–136.

Cheng, H. & T. Force. 2010. Molecular mechanisms of cardiovascular toxicity of targeted cancer therapeutics. *Circ. Res.* **106:** 21–34.

Clark, D.L., P.A. Andrews, *et al.* 1999. Predictive value of preclinical toxicology studies for platinum anticancer drugs. *Clin. Cancer Res.* **5:** 1161–1167.

Davila, J.C., R.J. Rodriguez, *et al.* 1998. Predictive value of *in vitro* model systems in toxicology. *Annu. Rev. Pharmacol. Toxicol.* **38:** 63–96.

Dickens, H., A. Ullrich, *et al.* 2008. Anticancer drug cis-4-hydroxy-L-proline: correlation of preclinical toxicology with clinical parameters of liver function. *Mol. Med. Report* **1:** 459–464.

Ehrich, M. 2003. Bridging the gap between *in vitro* and *in vivo* toxicology testing. *Altern. Lab. Anim.* **31:** 267–271.

Eschenhagen, T., T. Force, *et al.* 2011. Cardiovascular side effects of cancer therapies: a position statement from the Heart Failure Association of the European Society of Cardiology. *Eur. J. Heart Fail.* **13:** 1–10.

Fagerland, J. A., H. G. Wall, *et al.* 2012. Ultrastructural analysis in preclinical safety evaluation. *Toxicol Pathol.* **40:** 391–402.

Fielden, M.R., B.P. Eynon, *et al.* 2005. A gene expression signature that predicts the future onset of drug-induced renal tubular toxicity. *Toxicol. Pathol.* **33:** 675–683.

Fielden, M.R. & K.L. Kolaja. 2008. The role of early *in vivo* toxicity testing in drug discovery toxicology. *Expert Opin. Drug Saf.* **7:** 107–110.

Feldman, A.M., W.J. Koch, *et al.* 2007. Developing strategies to link basic cardiovascular sciences with clinical drug development: another opportunity for translational sciences. *Clin. Pharmacol. Ther.* **81:** 887–892.

Force, T., K. Kuida, *et al.* 2004. Inhibitors of protein kinase signaling pathways: emerging therapies for cardiovascular disease. *Circulation* **109:** 1196–1205.

Force, T., C.M. Pombo, *et al.* 1996. Stress-activated protein kinases in cardiovascular disease. *Circ. Res.* **78:** 947–953.

Force, T. & J.R. Woodgett. 2009. Unique and overlapping functions of GSK-3 isoforms in cell differentiation and proliferation and cardiovascular development. *J. Biol. Chem.* **284:** 9643–9647.

Fukumoto, J. & N. Kolliputi. 2012. Human lung on a chip: innovative approach for understanding disease processes and effective drug testing. *Front. Pharmacol.* **3:** 205.

Ganter, B., S. Tugendreich, *et al.* 2005. Development of a large-scale chemogenomics database to improve drug candidate selection and to understand mechanisms of chemical toxicity and action. *J. Biotechnol.* **119:** 219–244.

Geenen, S., P.N. Taylor, *et al.* 2012. Systems biology tools for toxicology. *Arch. Toxicol.* **86:** 1251–1271.

Higgins, J., M.E. Cartwright, *et al.* 2012. Progressing preclinical drug candidates: strategies on preclinical safety studies and the quest for adequate exposure. *Drug Discov. Today* **17:** 828–836.

Holsapple, M.P., C.A. Afshari, *et al.* 2009. Forum series: the "vision" for toxicity testing in the 21st century: promises and conundrums. *Toxicol. Sci.* **107:** 307–308.

Huh, D., D.C. Leslie, *et al.* 2012. A human disease model of drug toxicity-induced pulmonary edema in a lung-on-a-chip microdevice. *Sci. Transl. Med.* **4:** 159ra147.

Kim, H.J., D. Huh, *et al.* 2012. Human gut-on-a-chip inhabited by microbial flora that experiences intestinal peristalsis-like motions and flow. *Lab Chip* **12:** 2165–2174.

Kluwe, W.M. 1995. The complementary roles of *in vitro* and *in vivo* tests in genetic toxicology assessment. *Regul. Toxicol. Pharmacol.* **22:** 268–272.

MacDonald, J.S. & R.T. Robertson. 2009. Toxicity testing in the 21st century: a view from the pharmaceutical industry. *Toxicol. Sci.* **110:** 40–46.

Marchan, R., H.M. Bolt, *et al.* 2012. Systems biology meets toxicology. *Arch. Toxicol.* **86:** 1157–1158.

Marin, J.J., O. Briz, *et al.* 2009. Hepatobiliary transporters in the pharmacology and toxicology of anticancer drugs. *Front. Biosci.* **14:** 4257–4280.

Marks, L., S. Borland, *et al.* 2012. The role of the anaesthetised guinea-pig in the preclinical cardiac safety evaluation of drug candidate compounds. *Toxicol. Appl. Pharmacol.* **263:** 171–183.

Marx, U., H. Walles, *et al.* 2012. "Human-on-a-chip" developments: a translational cutting-edge alternative to systemic safety assessment and efficiency evaluation of substances in laboratory animals and man? *Altern. Lab Anim.* **40:** 235–257.

Meek, B. & J. Doull. 2009. Pragmatic challenges for the vision of toxicity testing in the 21st century in a regulatory context: another Ames test? . . . or a new edition of "the Red Book"? *Toxicol. Sci.* **108:** 19–21.

Mikaelian, I., M. Scicchitano, *et al.* 2013. Frontiers in preclinical safety biomarkers: microRNAs and messenger RNAs. *Toxicol. Pathol.* **41:** 18–31.

Olaharski, A.J., H. Uppal, *et al.* 2009. *In vitro* to *in vivo* concordance of a high throughput assay of bone marrow toxicity across a diverse set of drug candidates. *Toxicol. Lett.* **188:** 98–103.

Olson, H., G. Betton, *et al.* 2000. Concordance of the toxicity of pharmaceuticals in humans and in animals. *Regul. Toxicol. Pharmacol.* **32:** 56–67.

Pleil, J.D., M.A. Williams, *et al.* 2012. Chemical Safety for Sustainability (CSS): human *in vivo* biomonitoring data for complementing results from *in vitro* toxicology—-a commentary. *Toxicol. Lett.* **215:** 201–207.

Polson, A.G. & R.N. Fuji. 2012. The successes and limitations of preclinical studies in predicting the pharmacodynamics and safety of cell-surface-targeted biological agents in patients. *Br. J. Pharmacol.* **166:** 1600–1602.

Rennard, S.I., D.M. Daughton, *et al.* 1990. *In vivo* and *in vitro* methods for evaluating airways inflammation: implications for respiratory toxicology. *Toxicology* **60:** 5–14.

Robinson, J.F., P.T. Theunissen, *et al.* 2011. Comparison of MeHg-induced toxicogenomic responses across *in vivo* and *in vitro* models used in developmental toxicology. *Reprod. Toxicol.* **32:** 180–188.

Sai, K. & Y. Saito. 2011. Ethnic differences in the metabolism, toxicology and efficacy of three anticancer drugs. *Expert Opin. Drug Metab. Toxicol.* **7:** 967–988.

Warner, C.M., K.A. Gust, *et al.* 2012. A systems toxicology approach to elucidate the mechanisms involved in RDX species-specific sensitivity. *Environ. Sci. Technol.* **46:** 7790–7798.

Yang, B. & T. Papoian. 2012. Tyrosine kinase inhibitor (TKI)–induced cardiotoxicity: approaches to narrow the gaps between preclinical safety evaluation and clinical outcome. *J. Appl. Toxicol.* **32:** 945–951.

Ann. N.Y. Acad. Sci. ISSN 0077-8923

ANNALS OF THE NEW YORK ACADEMY OF SCIENCES
Issue: Annals *Meeting Reports*

Neuroprotection after cerebral ischemia

Shobu Namura,[1] Hiroaki Ooboshi,[2] Jialing Liu,[3] and Midori A. Yenari[4]

[1]Neuroscience Institute, Department of Neurobiology, Morehouse School of Medicine, Atlanta, Georgia. [2]Department of Internal Medicine, Fukuoka Dental College, Medical and Dental Hospital, Fukuoka, Japan. [3]Department of Neurological Surgery, [4]Department of Neurology, University of California, San Francisco, and The San Francisco Veterans Affairs Medical Center, San Francisco, California

Address for correspondence: Midori A. Yenari, M.D., Department of Neurology, University of California, San Francisco, and Department of Neurology, The San Francisco Veterans Affairs Medical Center, 4150 Clement St., San Francisco, CA 94121. yenari@alum.mit.edu

Cerebral ischemia, a focal or global insufficiency of blood flow to the brain, can arise through multiple mechanisms, including thrombosis and arterial hemorrhage. Ischemia is a major driver of stroke, one of the leading causes of morbidity and mortality worldwide. While the general etiology of cerebral ischemia and stroke has been known for some time, the conditions have only recently been considered treatable. This report describes current research in this field seeking to fully understand the pathomechanisms underlying stroke; to characterize the brain's intrinsic injury, survival, and repair mechanisms; to identify putative drug targets as well as cell-based therapies; and to optimize the delivery of therapeutic agents to the damaged cerebral tissue.

Keywords: stroke; cerebral ischemia; cerebrovascular disease; neurovascular unit; cell therapy; repair; immune response

Background and perspectives

Following interruption of blood flow to the brain in an ischemic event, cells undergo a series of events, such as loss of ion gradients, including failure of the sodium–potassium pump (Na–K), which leads to cellular swelling and cytotoxic edema. As cells switch from aerobic to anaerobic metabolism, metabolic acidosis ensues. Loss of ion gradients also leads to accumulation of intracellular calcium and excitatory amino acid (EAA) release, with activation of corresponding EAA receptors, leading to further calcium influx, mitochondrial dysfunction, and cell death through both necrotic and apoptotic pathways. Upon reperfusion, injured cells elicit a stress response, characterized by upregulation of immediate early and other stress response genes, which, in turn, leads to *in situ* production and/or upregulation of immune modulators, such as cytokines, and to trafficking of circulating immune cells into the ischemic brain. Necrotic cells may also release nucleic acids and other molecules that can act as damage-associated molecular patterns (DAMPs) on immune cells, including microglia, leading to immune cell activation and proinflammatory signaling. Proinflammatory molecules can then activate other proteins, such as matrix metalloproteinases (MMP), involved in the disruption of the blood–brain barrier (BBB) and the extracellular matrix. This worsens ischemic injury by causing vasogenic edema and hemorrhage. Reperfusion and immune cell signaling can also lead to astrocyte activation, with elaboration of prosurvival factors, setting the stage for reparative processes, such as neurogenesis and angiogenesis, as well as gliosis. The initial stress response also leads to induction of various survival factors, such as Akt and the cAMP response element–binding protein (CREB).

The "Trans-Pacific Workshop on Stroke" was held at the Wyndham Riverfront Hotel in New Orleans, Louisiana, on October 17–18, 2012, and was organized by Midori A. Yenari and Hiroaki Ooboshi, along with Shobu Namura and Jialing Liu. The workshop was sponsored by the U.S.–Japan Brain Research Cooperative Program and the Japan–U.S. Science and Technology Cooperation Program,

doi: 10.1111/nyas.12087

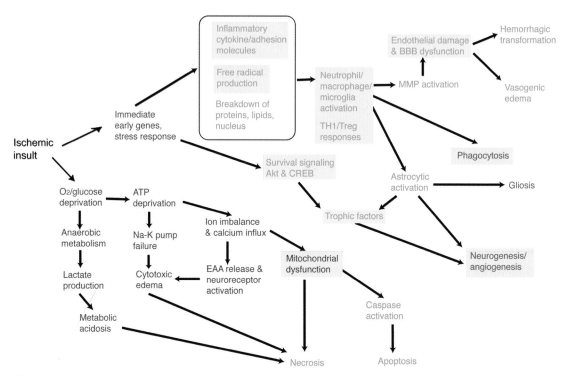

Figure 1. The ischemic cascade of events in the brain after ischemic stroke. Events are color coded according to their timing: red, acute phase (minutes to hours); green, subacute phase (hours to days); and blue, chronic phase (weeks to months). Cellular and molecular mechanisms that are highlighted in yellow were specifically featured by the workshop and discussed by the speakers.

Brain Research Division. Below is a detailed report of the workshop presentations and themes.[a]

Three goals were proposed to develop strong collaborations among investigators in both nations: (1) to exchange increasing knowledge of ischemic stroke, from the basic to clinic level, among researchers in the United States and Japan; (2) to identify barriers and gaps that inhibit complete understanding of the pathomechanisms underlying stroke; and (3) to identify research areas for future study. The scientific sessions were focused on several areas of investigation (Fig. 1): injury, survival, and repair mechanisms; potential drug targets and cell-based therapies; delivery of therapeutic agents, such as manipulating the BBB; and state-of-the-art imaging of neurovascular changes and for tracking deliv-

ered therapeutic agents. Speakers were selected from both countries to cover these areas. Participation of students and junior investigators was encouraged; emphasis was given to the diversification of the future workforce in stroke research, particularly in the United States. A total of 30 attendees were invited and presented their studies. A schematic of areas discussed as they relate to cell–cell interactions, signaling, and therapeutics within the neurovascular unit is shown in Figure 2.

Neuroprotection and clinical studies against ischemic stroke

The scientific session began with a talk by Kazuo Kitagawa (Osaka University), who has led the field of ischemic tolerance for more than two decades. He presented his recent progress in the understanding of the molecular mechanism underlying the phenomenon.[1] In the cytoplasm of neurons under nonstimulated conditions, SIK2 is highly expressed and phosphorylates and sequesters CRTC1. After ischemic stimuli, CaMK isoforms I/IV

[a]In this report, attendee presentations were organized thematically. Thus, the order of the report does not necessarily follow the order of the presentations. Discussions related to goals 2 and 3 are summarized below.

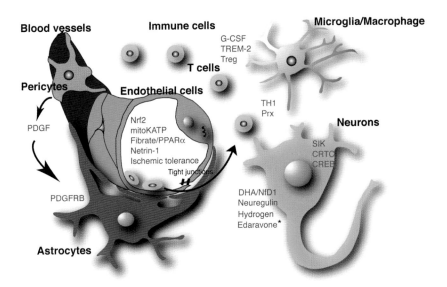

Figure 2. Schematic representation of cellular events in the neurovascular unit and potential therapeutic targets that were discussed at the workshop. Endothelial cells (blue) of the cerebral vessels and basal lamina (purple) are surrounded by an almost continuous layer of astrocyte (red) foot processes. Pericytes (orange) also cover the abluminal surface of the capillaries. In addition to these structures, tight junctions and transporters that are expressed on the endothelial cells contribute to the blood–brain barrier (BBB), a unique feature of cerebral vessels. After ischemia, in addition to neuronal injury, damage and activation of endothelial cells leads to BBB disruption and extravasation of blood-derived cells and serum molecules. Within the brain itself, microglia, the brain's resident immune cells, are activated. Endogenous molecules and responses, and protective agents are shown in blue text. Targets where inhibition is protective are shown in red text. (*, edaravone is already in use for treatment of ischemic stroke patients in Japan.)

phosphorylate SIK2, resulting in degradation of SIK2. Consequently, CRTC1 is dephosphorylated and translocates into the nucleus, binding to the promoter region of CREB. By doing so, this signaling pathway activates CREB-mediated survival genes, such as BDNF, PGC-1α, and Bcl2. Interestingly, this pathway (CaMK–SIK2–CRTC1–CREB) seems to be specifically downstream of synaptic NMDA receptor (NR2A-subunit containing) activation but not of other glutamate receptors. The concept was supported by *in vivo* findings in SIK2 null mice that showed strong resistance to ischemic stroke. Given the robustness of neuroprotection provided by this signaling pathway, further research should explore how to trigger this survival pathway pharmacologically, ideally after the onset of stroke.

Another neuronal survival mechanism was proposed by Shigeru Tanaka (Hiroshima University). Gene transfection of the G protein–coupled receptor (GPR)-3 to cultured neurons conferred resistance to hypoxia. By contrast, GPR3 gene knockdown by siRNA transfection augmented hypoxic neuronal apoptosis. Consistent with these findings, GRP3 knockout mice are more vulnera-

ble to transient focal cerebral ischemia compared with wild-type animals. GRP3 ligands may have a therapeutic value, although it is unknown how the GRP3 activation renders neuronal survival.

The role of autophagy, a cellular mechanism for clearing unnecessary debris, was discussed by Eisuku Dohi (Hiroshima University). He showed that disrupting chaperone-mediated autophagy worsened neuron death due to hypoxia.

The neuroprotective efficacies of docosahexaenoic acid (DHA), a fish oil component, and its derivative neuroprotection D1 (NPD1), were shown by investigators at Louisiana State University. Ludmila Belayev tested post-stroke intravenous injection of DHA (22:6n–3) in rats subjected to two hours of middle cerebral artery occlusion (MCAO).[2] DHA significantly improved behavior outcomes and attenuated brain edema (MRI T2-weighted image) and infarct size up to seven days after stroke. Such improvements were seen when DHA was injected at five hours after onset of stroke (i.e., three hours after reperfusion). Enhanced NPD1 synthesis in the brain penumbra area in DHA-treated animals was demonstrated by lipidomic analysis.

Nicolas Bazan (Louisiana State University) presented his unique approach using the combination of aspirin and DHA. Aspirin alone has been shown to afford beneficial effects against cerebrovascular diseases. Bazan found that aspirin and DHA cotreatment induced the synthesis of aspirin-triggered NPD1 (AT-NPD1) in the brain.[3] Injection of AT-NPD1 sodium salt or methyl ester at three hours after stroke onset was effective in improving outcomes in rats subjected to two hours of MCAO.

Alberto Musto (Louisiana State University) reported protection by NPD1 in a mouse status epilepticus model induced by pilocarpine. Post-status epilepticus NPD1 treatment reduced recurrent seizure frequency and improved electrophysiological outcomes, suggesting that neuroprotection by NPD1 could operate at the postsynaptic level.[4]

Another potential neuroprotectant candidate was proposed by Byron Ford (Morehouse School of Medicine). Ford studied neuregulin-1, which had originally been identified as a growth factor at the neuromuscular junction. Carotid arterial injection with neuregulin-1 showed impressive infarct reductions and neurological improvements in rats.[5] Zhenfeng Xu, a member of Ford's laboratory, presented their recent attempt at genomic and transcriptomic approaches using rat brain samples after stroke. Ford also discussed his recent experience with a nonhuman primate model in searching for potential biomarkers for predicting stroke outcomes.

Mami Noda (Kyushu University) discussed her unique neuroprotection approach using molecular hydrogen. She tested the oral administration of hydrogen-containing water in rodent models of Parkinson's disease[6] and in an ischemic model of the optic nerve. Although the hydrogen level in the brain was not influenced, and although the underlying mechanism remained to be studied, drinking hydrogen-containing water was protective against those pathological conditions in these models. Since inhalation of hydrogen gas has been shown to protect against ischemic stroke in rats, drinking hydrogen-containing water may also protect against stroke.

Findings from recent clinical studies were also presented. Shunya Takizawa (Tokai University) reported the outcomes of a phase I study of intravenous granulocyte colony-stimulating factor (G-CSF) in ischemic stroke patients.[7] The clinical study was based on the previous findings of Takizawa and colleagues, in mice, that hematopoietic cytokines reduced infarct volume with improvements in motor and cognitive functions. According to the phase I study, G-CSF (150 and 300 μg/day) was safe and well tolerated in patients after ischemic stroke. A phase II study is underway testing G-CSF treatment 24 hours after onset.

Koji Abe (Okayama University) presented the outcomes of edaravone (Radicut®) in a retrospective study of 114 consecutive stroke patients who received tissue plasminogen activator (tPA, 0.6 mg/kg) within three hours after onset.[8] Edaravone is a free radical scavenger that was approved for treating acute ischemic stroke in Japan in 2001. Edaravone-treated patients showed a higher recanalization rate after tPA compared to those who did not receive edaravone, with the caveat that edaravone-treated patients had a higher prevalence of cardiogenic embolism and lower NIHSS scores on admission.

Cerebrovasculature and neurovascular protection against ischemic stroke

Study of the cerebrovasculature and its interaction with the neurovascular unit are traditional and indispensable areas confronting stroke research. Recent advances in the pathophysiology of the BBB and novel approaches to cerebrovascular regulation were discussed.

Tetsuro Ago (Kyushu University) discussed the neurotrophic roles of pericytes. He showed that the expression of platelet-derived growth factor receptor β (PDGFRβ) is drastically elevated in pericytes of the peri-infarct areas in mice after MCAO.[9] Heterozygous PDGFRβ gene knockout suppressed the elevation of PDGFRβ with increased infarct formation. In addition, PDGFβ, the ligand for PDGFRβ, elevates the expression of nerve-growth factor (NGF) and neurotrophin-3 (NT-3) in cultured pericytes. These findings suggest that the PDGFβ-mediated NGF/NT-3 production from pericytes may afford neuroprotection against ischemic stroke.

David S. Miller (National Institute of Environmental Health Science) described the importance of nuclear factor E2–related factor 2 (Nrf2) in drug delivery through the BBB. Nrf2 is a redox-sensitive, ligand-activated transcription factor that induces multiple antioxidant and glutathione-generating enzymes in response to oxidative stress. He assessed the effects of Nrf2 activation at the BBB on drug efflux transporters, including P-glycoprotein.[10] The Nrf2 ligand sulforaphane elevated protein expression of P-glycoprotein as well as the efflux activity in isolated rodent brain capillaries, suggesting that this pathway contributes to restricting drug delivery across the BBB when Nrf2 is activated.

Jeffrey M. Gidday (Washington University) highlighted a discussion of vascular mechanisms involved in ischemic tolerance. He demonstrated that hypoxic preconditioning reduced post-stroke leukocyte adhesion and BBB dysfunction. Downregulation of intercellular adhesion molecules, as well as enhanced integrity of tight junctions (e.g., ZO-1 and claudin-5), has been documented as a common phenotype in vascular ischemic tolerance.[11] A better understanding of the vascular aspects of ischemic tolerance is warranted.

David Busija (Tulane University) summarized his research regarding vascular mitochondria as a therapeutic targets. Transient activation of ATP-dependent potassium channels that are presented on the inner mitochondrial membrane (mitoKATP channels) induces immediate and long-term protection of the cerebral endothelium against subsequent stress.[12] Attenuations in both intracellular calcium elevation and reactive oxygen species production after stress are likely to induce this protection. In addition, many of the molecular consequences of mitoKATP channel activation have the potential to influence cerebrovascular tone. These phenomena are blunted in insulin-resistant Zucker obese rats, suggesting negative impacts of abnormal glucose/lipid metabolism, common comorbidities in stroke patients. Prasad V.G. Katakam (Tulane University) showed that mitochondrial activation also promotes neuronal isotype-mediated nitric oxide (NO) generation, suggesting the existence of a novel link between neuronal metabolism and vasodilation.

A therapeutic approach targeting peroxisome proliferator–activated receptor (PPAR) α in cerebral vessels was discussed by Shobu Namura and Donghui Li (Morehouse School of Medicine). The PPARs are nuclear receptors that act as transcription factor. Fibrates, clinically used drugs for dyslipidemia, are known to activate PPARα. Fibrates improve cerebral blood flow (CBF) in the penumbral area.[13] Fibrate-induced elevation of superoxide dismutase activity in brain microvessels may contribute to maintaining NO bioavailability, with improvement of ischemic cerebral blood flow.

Immune responses and ischemic stroke

Immune responses following stroke continue to be an active area of investigation, and both innate and adaptive immune responses in stroke were discussed. Innate immune responses have been more extensively investigated, since acute neurological insults were not traditionally considered in the context of prior antigen exposure. Newer aspects of innate immunity were also discussed, as they relate to acute and long-term effects of stroke. Adaptive immunity may be relevant in the search for a vaccine against stroke and may also help to explain why concurrent infections are detrimental to stroke outcome.

Kyra Becker (University of Washington) reviewed the literature on adaptive immune responses in experimental stroke and presented a new model of adaptive immunity, whereby the systemic administration of lipopolysaccharide at the time of stroke led to Th1 responses and worsened outcome.[14] This model could be likened to the negative outcomes in stroke patients with complicating infections. Interestingly, adoptive transfer of splenocytes primed toward a Th1 response led to a worsened outcome, whereas adoptive transfer of splenocytes primed toward a T_{reg} cell response led to a better outcome. These data suggest that interventions preventing Th1 or enhancing T_{reg} cell responses may have translational value.

A study of novel immune molecules in experimental stroke was presented by Hiroaki Ooboshi (Fukuoka Dental College). Prior work has shown that lymphocytes contribute to adverse outcomes. However, the role of T lymphocyte subtypes has not been well studied. The γδ T cells can produce IL-17 following stimulation by macrophage-derived IL-23. These cytokines appear to contribute to stroke evolution, as mice deficient in these cytokines are protected.[15] Further, peroxiredoxin (Prx), an endogenous antioxidant, induces IL-23 and leads to worsened stroke outcome through Toll-like receptors-2 and -4 and the MyD88 pathway. These findings show that Prx is a novel damage-associated molecular signal. Inhibiting Prx appears to be protective.

While many of the immune responses described to date are largely detrimental in the acute phase, Midori A. Yenari (University of California, San Francisco) described a newly characterized innate immune receptor that has a potentially beneficial role. This receptor, triggering receptors expressed on myeloid cells-2 (TREM-2), is thought to trigger phagocytosis in microglia and macrophages. Its ligand has been identified in brain cells, and recent work has shown that exposure of neurons to apoptotic insults leads to the activation of TREM-2.[16] TREM-2 deficiency decreased phagocytosis of injured neurons. Masahito Kawabori (University of California, San Francisco) then showed that the proportion of microglia with TREM-2 expression appears to be enhanced under conditions of therapeutic hypothermia. Thus, TREM-2 may contribute beneficial effects, such as the clearance of cellular debris.

The link between pain and inflammation is well known, and Nozomi Akimoto (Kyushu University) presented new findings showing how the CCL-1 cytokine enhances nociception and microglial activation.

Recovery and repair, and modeling of post-stroke fatigue

Since the discovery of endogenous neural stem cells in the adult brain, and following recent reports that post-stroke systemic administration of mesenchymal stem cells improved outcomes in animals, neuronal repair has rapidly emerged as a potential method for treating stroke. One week before the workshop, the Nobel Prize in Physiology or Medicine 2012 was awarded to Sir John B. Gurdon and Shinya Yamanaka for the discovery that mature cells can be reprogrammed to pluripotency. Presentations concerning this topic were actively discussed.

Koji Abe (Okayama University) discussed his exploration of the potential of the induced pluripotent stem (iPS) cell transplantation as a novel therapy for ischemic stroke. Unexpectedly, intracranially transplanted iPS cells formed teratomatous tumors in the ischemic mouse brains and the clinical recovery from stroke was delayed, despite increases in the number of neuroblasts and mature neurons in the ischemic brains.[17] He concluded that iPS cell therapy had a promising potential to provide neural cells after stroke if tumorigenesis could be controlled.

Optimal delivery methods and timing of neural stem cell therapy against stroke were discussed by Raphael Guzman (University Hospital Basel).[b] He recommended that intravascular injection was advantageous over intraparenchymal transplantation for achieving a widespread distribution while being minimally invasive and repeatable. Compared to intravenous injection, which often results in entrapment of transplanted cells in the lung, intra-arterial injection provides better homing into the brain. With reference to timing of the intra-arterial approach, injection at three days after ischemia resulted in the highest cell engraftment.[18] Pretreatment of transplanted cells with BDNF enhanced the therapeutic efficacies.

Another recovery approach targeting angiogenic factor Netrin-1 was discussed by Jialing Liu and Chin Cheng Le (University of California, San Francisco). Netrin-1 gene delivery into the ischemic penumbra not only increased vascular density but also promoted the migration of immature neurons into the peri-infarct white matter, which was accompanied by improved recovery of motor function.[19] The enhanced neurogenesis by Netrin-1 likely contributes to neurological recovery, because conditional ablation of neuroprogenitor cells through targeting nestin in adult mice delayed the recovery of cognitive function after stroke without affecting CBF and subsequent lesion size.

Finally, Allison Kunze (University of Washington) presented a novel approach for detecting post-stroke fatigue in popular rodent stroke models. This approach should open up the field to the identification of new treatments for a significant but understudied clinical problem.

Novel imaging technologies for stroke studies

In vivo optic brain imaging is an emerging area in experimental stroke studies. For instance, as the fluorescence labeling technique advances, two photon–excited microscopy provides enormous potential for repeated documentation not only of neurovascular morphology and hemodynamics, but also of changes at the molecular level, including those in intracellular ion levels.

---

[b]At the time of the conference, Dr. Guzman was affiliated with Stanford University.

Chris Schaffer (Cornell University) discussed his experience in developing microscale stroke models utilizing femtosecond laser pulses.[20] By controlling the amount of delivered energy, either vessel occlusion or rupture can be produced at a precise point. With regard to neocortex microinfarct, the location of the occlusion determines the consequence: in contrast to the sustained perfusion deficit in the distal portions after occlusion of a penetrating arteriole, there was a robust reversal flow from distal branches after occlusion of cortical surface arterioles. On the other hand, rupture of a penetrating arteriole (microhemorrhage) triggered local inflammatory responses without detectable neurovascular pathology, such as dendrite deformation and capillary collapse, in the surrounding tissue. Nozomi Nishimura (Cornell University) showed in a mouse model of Alzheimer's disease that capillary flow was dramatically stalled, which was accompanied by leukocyte adhesion to the endothelium. The capillary plugging by leukocytes may cause CBF impairment in Alzheimer's disease.

Jialing Liu and Yosuke Akamatsu (University of California, San Francisco) shared their experience with optical coherence tomography and optical microangiography for measuring cerebrovascular microstructure and flow, and post-ischemia collateral circulation in a mouse model of type 2 diabetes mellitus (*db/db*). Type 2 diabetes mellitus is known to be associated with worse stroke outcomes. Compared with nondiabetic *db/+* mice, *db/db* showed lower regional CBF and lower density of functional blood vessels in the ischemic hemisphere. These new imaging methods are useful for monitoring collateral flow development in mice after stroke.

Overall summary and future directions

The meeting was viewed by most of the attendees as a successful beginning to the development of collaborative efforts between the two countries in the investigation of the pathomechanisms of ischemic stroke. The size and format of the meeting were well received, especially by the Japanese participants and junior investigators, who often feel intimidated when asked to speak at large conferences. The location and timing surrounding the Society for Neuroscience meeting was convenient for the U.S. participants, particularly the basic researchers and students.

This meeting clearly demonstrated the need for future meetings that could expand upon the topics covered. For example, edaravone has been used with stroke patients in Japan, and U.S. participants expressed an interest in learning more about the experiences of Japanese clinicians, with an eye toward larger-scale international studies. Neural repair using iPS was another actively discussed topic. Because of its accessibility, pluripotency, and autologous nature, iPS technology has tremendous potential for treating neurodegenerative diseases, including stroke. Future investigations that apply iPS technology to the treatment of stroke patients may be an important area of collaboration between investigators in the two countries. Establishing standardized preclinical models and unbiased study methods would allow comparison of biological robustness of findings across laboratories, which is important when considering clinical translation. Related to this issue, developing a reproducible non-human primate model of stroke will certainly be useful for testing promising neuroprotectants, such as neuregulin-1, DHA, and NPD1. Collaborations with physicists and chemists are needed, as shown by the examples of *in vivo* imaging.

Acknowledgments

This meeting was funded by an NIH R13 Grant through the U.S.–Japan Brain Research Cooperative Program (3R01 NS040516-11S1) to Midori A. Yenari; and from the Japan–U.S. Science and Technology Cooperation Program, Brain Research Division, to Hiroaki Ooboshi, Genentech (a member of the Roche Group), and Peter R. MacLeish (Neuroscience Institute, Morehouse School of Medicine). Grant funds were administered by the Northern California Institute for Research and Education. The organizers of the "Trans-Pacific Workshop on Stroke" thank the presenters and attendees for their contributions and lively and insightful discussions. The attendees dedicated the meeting to the memory of Akira Arimura, an American scientist at Tulane University with beginnings in Japan, whose remarkable career included many U.S.–Japan collaborations.

Koji Abe gave the featured lecture, "Current topics on neuroprotection after cerebral ischemia," which included a brief report of the impact of the earthquake/tsunami in March 2011 on healthcare service in the affected Tohoku pacific coastal areas of Japan. His report reminded the attendees that the workshop was located in the area hit by Hurricane Katrina seven years before. Local speakers Nicolas Bazan, Ludmila Belayev, and David Busija were featured.

Conflicts of interest

The authors have no conflicts of interest.

References

1. Sasaki, T., H. Takemori, Y. Yagita, *et al.* 2011. SIK2 is a key regulator for neuronal survival after ischemia via TORC1-CREB. *Neuron* **13:** 106–119.
2. Belayev, L., L. Khoutorova, K.D. Atkins, *et al.* 2011. Docosahexaenoic acid therapy of experimental ischemic stroke. *Transl. Stroke Res.* **2:** 33–41.
3. Bazan, N.G., T.N. Eady, L. Khoutorova, *et al.* 2012. Novel aspirin-triggered neuroprotectin D1 attenuates cerebral ischemic injury after experimental stroke. *Exp. Neurol.* **236:** 122–130.
4. Musto, A.E., P. Gjorstrup & N.G. Bazan. 2011. The omega-3 fatty acid-derived neuroprotectin D1 limits hippocampal hyperexcitability and seizure susceptibility in kindling epileptogenesis. *Epilepsia* **52:** 1601–1608.
5. Xu, Z., D.R. Croslan, A.E. Harris, *et al.* 2006. Extended therapeutic window and functional recovery after intraarterial administration of neuregulin-1 after focal ischemic stroke. *J. Cereb. Blood Flow Metab.* **26:** 527–535.
6. Fujita, K., T. Seike, N. Yutsudo, *et al.* 2009. Hydrogen in drinking water reduces dopaminergic neuronal loss in the 1-methyl-4-phenyl-1,2,3,6-tetrahydropyridine mouse model of Parkinson's disease. *PLoS One* **30:** e7247.
7. Moriya, Y., A. Mizuma, T. Uesugi, *et al.* 2012. Phase I Study of intravenous low-dose granulocyte colony-stimulating factor in acute and subacute ischemic stroke. *J. Stroke Cerebrovasc. Dis.* DOI: 10.1016/j.jstrokecerebrovasdis.2012.08.002.
8. Kono, S., K. Deguchi, N. Morimoto, *et al.* 2011. Tissue plasminogen activator thrombolytic Therapy for acute ischemic stroke in 4 hospital groups in Japan. *J. Stroke Cerebrovasc. Dis.* DOI: 10.1016/j.jstrokecerebrovasdis.2011.07.016.
9. Arimura, K., T. Ago, M. Kamouchi, *et al.* 2012. PDGF receptor β signaling in pericytes following ischemic brain injury. *Curr. Neurovasc. Res.* **9:** 1–9.
10. Miller, D.S. 2010. Regulation of P-glycoprotein and other ABC drug transporters at the blood-brain barrier. *Trends. Pharmacol. Sci.* **31:** 246–254.
11. Wacker, B.K., A.B. Freie, J.L. Perfater & J.M. Gidday. 2012. Junctional protein regulation by sphingosine kinase 2 contributes to blood-brain barrier protection in hypoxic preconditioning-induced cerebral ischemic tolerance. *J. Cereb. Blood Flow Metab.* **32:** 1014–1023.
12. Katakam, P.V., F. Domoki, J.A. Snipes, *et al.* 2009. Impaired mitochondria-dependent vasodilation in cerebral arteries of Zucker obese rats with insulin resistance. *Am. J. Physiol. Regul. Integr. Comp. Physiol.* **296:** R289–298.
13. Guo, Q., G. Wang & S. Namura. 2010. Fenofibrate improves cerebral blood flow after middle cerebral artery occlusion in mice. *J. Cereb. Blood Flow Metab.* **30:** 70–78.
14. Zierath, D., J. Hadwin, A. Savos, *et al.* 2010. Anamnestic recall of stroke-related deficits: an animal model. *Stroke* **41:** 2653–2660.
15. Shichita, T., E. Hasegawa, A. Kimura, *et al.* 2012. Peroxiredoxin family proteins are key initiators of post-ischemic inflammation in the brain. *Nat. Med.* **18:** 911–917.
16. Hsieh, C.L., M. Koike, S.C. Spusta, *et al.* 2009. A role for TREM2 ligands in the phagocytosis of apoptotic neuronal cells by microglia. *J. Neurochem.* **109:** 1144–1156.
17. Kawai, H., T. Yamashita, Y. Ohta, *et al.* 2010. Tridermal tumorigenesis of induced pluripotent stem cells transplanted in ischemic brain. *J. Cereb. Blood Flow Metab.* **30:** 1487–1493.
18. Rosenblum, S., N. Wang, T.N. Smith, *et al.* 2012. Timing of intra-arterial neural stem cell transplantation after hypoxia-ischemia influences cell engraftment, survival, and differentiation. *Stroke* **43:** 1624–1631.
19. Sun, H., T. Le, T.T.J. Chang, *et al.* 2011. AAV-mediated netrin-1 overexpression increases peri-infarct blood vessel density and improves motor function recovery after experimental stroke. *Neurobiol. Dis.* **44:** 73–83.
20. Nishimura, N., C.B. Schaffer, B. Friedman, *et al.* 2006. Targeted insult to subsurface cortical blood vessels using ultrashort laser pulses: three models of stroke. *Nat. Methods* **3:** 99–108.

Ann. N.Y. Acad. Sci. ISSN 0077-8923

ANNALS OF THE NEW YORK ACADEMY OF SCIENCES

Reviewers for *Annals of the New York Academy of Sciences*, 2011–2012

A

Abel, Laurent
Abhyankar, Avinash
Adolphs, Ralph
Agodi, Antonella
Ahima, Rexford
Ahrén, Bo
Akashi, Koichi
Alcaraz, Daniel
Alexandrov, Andrei
Alou-Chebl, Alex
Alvarez, Francisco
Amasheh, Maren
Amasheh, Salah
Amunts, Katrin
Anderson, James
Andersson, Leif
Angata, Takashi
Angert, Amy
Arai, Fumio
Arnoult, Damien
Arregui, Inigo
Aschenbach, Jörg
Atkinson, Mark
Auerbach, Benjamin
Awasthi, Amit
Aziz, Tipu

B

Bahr, Janice
Balasubramanyam, Ashok
Banks, William
Barbash, Dan
Basu, Joyoti
Bauer, Margaret
Bauer, Thomas
Baum, Christopher
Baum, Linda
Bax, Ad
Bayer, Arnold
Beaton, Alan
Beeson, David

Benoit, Stephen
Bernat, James
Berrih-Aknin, Sonia
Beyenbach, Klaus
Bian, Weining
Bickel, Perry
Bijlsma, Rudolf
Binder, Henry
Birnbaum, Morris
Black, Lauren
Blasig, Ingolf
Bleich, Markus
Bock-Marquette, Ildiko
Bolam, John Paul
Bonomo, Robert
Borden, William
Boucher, Helen
Boutin, Stanley
Bowyer, Terry
Bradford, Patricia
Brandner, Johanna
Bratton, Donna
Bray, George
Brehm, Michael
Brewer, Curtis
Brickner, Steven
Brissova, Marcela
Brown, Donald
Bruce, Laura L.
Brunham, Robert
Bucher, Doris
Bücker, Roland
Buckley, Lauren
Buettner, Christoph
Bujalska, Iwona
Bulter, Andrew
Burns, Ted

C

Cadenas, Enrique
Cain, Barak
Calabrese, Vittorio

doi: 10.1111/nyas.12099
Ann. N.Y. Acad. Sci. 1278 (2013) 33–39 © 2013 New York Academy of Sciences.

Calado, Rodrigo
Cantalupo, Claudio
Cao, Xu
Cao, Xuetao
Caplan, Louis R.
Carr, Rotonya
Carter, Anthony
Casanova, Jean-Laurent
Cashmore, Lisa
Castagnola, Massimo
Chapel, Helen
Chapin, Terry
Chaves, Luis
Chen, Di
Chini, Eduardo
Christadoss, Premkumar
Christensen, Norm
Chrousos, George
Chu, Wen-Ming
Ciafaloni, Emma
Citi, Sandra
Civitelli, Roberto
Coast, Geoffrey
Cobb, Brian
Colgan, John
Collins, Shelia
Colman, Peter
Conant, Gavin
Conley, Mary Ellen
Conner, Mary
Cools, Roshan
Cooper, Dianne
Cooper, Leslie
Cooper, Max
Coppari, Roberto
Corbetta, Daniela
Cormode, Graham
Coulie, Pierre
Cowley, Kristine
Crocker, Paul
Crow, Timothy
Crozier, Alan
Cui, Long
Cundy, Tim
Cupedo, Tom

D

D'Alessio, David
Dahlman-Wright, Karin

Danese, Silvio
Daniels, Stephen
Darley-Usmar, Victor
Davidson, Alan J.
De Baets, Marc
de Ferranti, Sarah
De Groot, Anne
de Laat, Bas
de Pedro, Miguel
de Villartay, Jean-Pierre
de Waal Malefyt, Rene
DeAngelis, Don
DeBenedetto, Anna
del Zoppo, Greg
Demetriou, Michael
DeVault, Travis
Diederich, Marc
DiMaio, J. Michael
Dirnagl, Ulrich
Dixit, Vishwa
Donath, Marc
Douthwaite, Stephen
Duan, Wenzhen
Duckworth, Renee
Duffy, Erin
Duffy, Meghan
Dushay, Jody

E

Easton, J. Donald
Eaves, Connie
Edenberg, Howard
Eggeling, Christian
Ehrlich, H
Einsele, Hermann
Engel, Andrew
Epple, Hans-Joerg
Erev, Ido
Escobar, Carolina
Estes, James
Evoli, Amelia
Evora, Paulo

F

Faa, Gavino
Faith, Myles
Fall, Tove
Fallon, Padraic
Fanning, Alan

Farrugia, Maria Elena
Fasano, Alessio
Fasano, Alfanso
Faulconbridge, Lucy
Feizi, Ten
Fennoy, Ilene
Fernandes, Prabha
Fiaschi-Taesch, Nathalie
Figdor, Carl
Filipovich, Alexandra
Fimognari, Carmela
Fischer, Alain
Fisher, Mark
Fitch, W. Tecumseh
Fitzgerald, Maria
Fletcher, Alison
Foster, Susan
Frank, James
Franke, Werner
Frantz, Stefan
Freund, Alexandra
Frumkin, Howard
Fukuda, Minoru
Fuleihan, Ramsay
Furuse, Mikio

G
Gabius, Hans-Joachim
Gaglia, Jason
Galgani, Jose
Gapin, Laurent
Gardi, Concetta
Geber, Monica
Geha, Raif
Genesee, Fred
Gennery, Andrew
Ghose, Subroto
Gibbs, James
Gibran, Nicole
Gibson, Peter
Gilhus, Nils Erik
Gilkeson, Gary
Giovanetti, Anne
Giszter, Simon
Gomez-Pinilla, Fernando
Gonder, Katy
González-Mariscal, Lorenza
Gordon, Joshua
Gorelick, David

Gozal, David
Graber, Alan
Graf, Peter
Green, Jeffrey
Greening, Andrew
Grimson, Andrew
Groothuis, Ton
Gross, James
Grotta, James
Günzel, Dorothee
Gupta, Sudhiranjan
Guptill, Jeffrey

H
Hadjivassiliou, Marios
Haisch, Lea
Handgretinger, Rupert
Hannappel, Ewald
Haq, Ihtsham
Harnett, William
Harper, Charles
Harren, Frans
Hart, Patrick
Harte, John
Haseloff, Reiner
Hayasaka, Satoru
Hayday, Adrian
Haynes, Laura
Heath, Matthew
Hecht, Gail
Heiss, Wolf-Dieter
Hematti, Peiman
Heng, Ik Siong
Hennessy, Michael
Hering, Nina
Hiepe, Falk
Higgins, Steve
Hirabayashi, Jun
Hochman, Shawn
Hokke, Cornelis
Holland, Steven
Holst, Jens
Hopkins, William
Horwitz, Greg
Hou, Jianghui
Houghton, Richard
Houle, John
Huang, Christopher
Huber, Otmar

Hursting, Stephen
Husebye, Eystein

I

Imai, Yumi
Iqbal, Jameel
Irwin, Rebecca
Ivashkiv, Lionel
Iwama, Atsushi

J

Jackson, John
Jackson, Stephen
Jagasia, Shubhada
Jaswinder, Sethi
Jernigan, Terry
Jimenez, Ivan
Johnson, Ed
Johnson, Eric
Johnson, Pauline
Jones, H.
Jongejans, Eelke
Jung, Ranu
Junko, Shigemitsu

K

Kadereit, Suzanne
Kalsbeek, Andries
Kamel, Hooman
Kaminski, Clemens
Karlsson, Stefan
Katiyar, Santosh
Kay, Lewis
Keane, Bob
Kellner, Jim
Kenny, Paul
Kensinger, Elizabeth
Kent, Sally
Kent, Thomas
Kincade, Paul
Kivisild, Toomas
Klar, Amar
Kleinman, Hynda
Kojima, Takashi
Kokko, Hanna
Kondoh, Masuo
Konrad, Martin
Koob, George
Korkina, Liudmila

Kotz, Catherine
Kousteni, Stavroula
Koval, Michael
Krause, Gerd
Krejsek, Jan
Krug, Susanne
Kuitunen, Katja
Kullander, Klas
Kupatt, Christian
Kvarnemo, Lotta

L

Lambert, Paul-Henri
Landi, Enrico
Landry, Gregory
Lansdorp, Peter
Lau, Joseph
Leahy, Jack
Lebwohl, Mark
Leikauf, George
Lengerke, Claudia
Lenz, Frederick
Leopold, Donald
Lesage, Jean
Levine, James
Levine, Jon
Liang, Guosheng
Lindholm, Dan
Lindstrom, Jon
Lisak, Robert
Liu, Fu-Tong
Liu, Tau
Lohoff, Michael
London, Edythe
Lotze, Heike
Lövblad, Karl-Olof
Lubman, David
Luparello, Claudio
Luxon, Bruce

M

MacDermott, Amy
Macfadyen, Sarina
Maddison, Paul
Mai, Jurgen
Majeti, Ravindra
Mankin, Alexander
Mantegazza, Renato
Manz, Markus

Marks, William
Marx, Alexander
Maselli, Ricardo
May, James
Mayberg, Helen
McCarthy, Margaret
McClung, Colleen
McCole, Declan
McCormick, Catherine
McGinnis, William
McGrew, Michael
McGrew, William
McManus, Chris
Meddings, Jon
Mehta, Anand
Meirhaeghe, Aline
Melander, Christian
Meltzer, Peter
Meriggioli, Matthew
Meriney, Stephen
Meroni, Pier Luigi
Merrill, Joan
Mestecky, Jiri
Milatz, Susanne
Miller, Marvin
Minegishi, Yoshiyuki
Mirmira, Raghu
Mitscher, Lester
Moczek, Armin
Monecke, Stefan
Montoya, Ivan
Moore, Dan
Moore, Michael
Moore, Peter
Morley, John
Mosoian, Arevik
Mudway, Ian
Müller, Gerd
Munafò, Marcus
Mutschler, Isabella

N
Nagarajan, Sri
Narimatsu, Hisashi
Naylor, Paul
Naz, Rajesh
Newman, Amy
Nguyen, Kytai

Nillni, Eduardo
Nimmerjahn, Falk
Nishimune, Hiroshi
Noel, Gary
Notarangelo, Luigi
Novak, Colleen
Nunney, Len
Nusrat, Asma

O
O'Brien, Charles
O'Connor, Kevin
O'Doherty, John
O'Donovan, Michael
Olivola, Christopher
Olivot, Jean-Marc
Ollero, Mario
Oostendorp, Robert
Opferman, Joseph
O'Shea, John
Ouyang, Weiming

P
Pacifici, Roberto
Padoa-Schioppa, Camillo
Page, Malcolm
Palzkill, Tim
Pan, Weihong
Pan-Hammartström, Qiang
Parker, Patty
Parkos, Charles
Parr, Cynthia
Parr, Thomas
Pascual, Virginia
Pastor, John
Paulson, James
Pavlidis, Paul
Pearlson, Godfrey
Pecorelli, Alessandra
Perretti, Mauro
Pfaff, Samuel
Philips, Jennifer
Picard, Capucine
Pierce, Michael
Pimm, Stuart
Pomba, Constanca
Popadic, Aleksandar
Poritz, Lisa
Poulin-Dubois, Diane

Powers, Al
Powers, Alice
Prochazka, Arthur
Projan, Steve
Purves, Dale
Purvis, Andy

Q
Qvarnstrom, Anna

R
Rabinovich, Gabriel
Raeyemaekers, Joost
Rajput, Akhil
Ramirez, Nino
Rauvala, Heikki
Raymond, Michel
Read, Tim
Reddel, Stephen
Renowden, Shelly
Reyna, Valerie
Rezaei, Nima
Reznick, Abraham
Richard, Byrne
Richman, David
Ripperger, Juergen
Rittner, Heike
Robak, Tadeusz
Robbins, Trevor
Rogler, Gerhard
Romani, Luigina
Rose, Jed
Rosenthal, Rita
Rosi-Marshall, Emma
Rossignol, Serge
Rostami, Abdolmohamad
Rousseff, Rossen
Routes, John
Rucker, Janet
Ruddle, Nancy
Runge, Michael
Rybak, Jürgen

S
Sahl, Hans-Georg
Sakai, Randall
Sakurai, Takuya
Salvatori, Giovanni
Samanez-Larkin, Gregory

Sawado, Tomoyuki
Schacter, Daniel
Scharl, Michael
Schiff, Nicholas
Schlaepfer, Thomas
Schluter, Dolph
Schmidt-Ott, Kai
Schreiber, John
Schumann, Michael
Schwartz, Michael
Seddon, Phillip
Seidler, Rachael
Serviddio, Gaetano
Sesack, Susan
Sette, Alessandro
Seyfried, Salim
Shaw, Jonathan
Shearer, William
Shen, Le
Sherwood, Chet
Shi, Yufang
Shinabarger, Dean
Shirtliff, Mark
Silverman, Jared
Simpson, Evan
Šimúth, Jozef
Sleep, Norman
Smaers, Jeroen
Smart, Nicola
Snyder, Solomon
Söderholm, Johan
Sosne, Gabriel
Spaink, Herman
Sparling, Frederick
Sperandio, Markus
Speth, Gus
Spiller, David
Srivastava, Deepak
St. Jacques, Peggy
Staal, Frank
Stadecker, Miguel
Staels, Bart
Stanford, Arielle
Steele, Vernon
Stein, Jürgen
Stein, Paul
Stevenson, Richard
Stranges, Saverio

Ann. N.Y. Acad. Sci. 1278 (2013) 33–39 © 2013 New York Academy of Sciences.

Strober, Warren
Suman, Paritosh
Sun, Herb
Sun, Li
Surh, Young-Joon
Sutcliffe, Joyce
Sutherland, Ann
Swann, Gary
Swanson, David

T
Tait, Brian
Takayanagi, Hiroshi
Talbot, George
Tang, Bor Luen
Tangye, Stuart
Teitelbaum, Daniel
Thomas, David
Thomas, James
Thornton, Wendy
Titulaer, Maarten
Togari, Akifumi
Toma, Ian
Tomaras, Andrew
Tonks, Amanda
Tononi, Giulio
Toscano, Marta
Treadwell, Terry
Tsubura, Airo
Tsukita, Sachiko
Turner, Jerrold
Tuthill, Cynthia
Tutkuviene, Janina

U
Uhl, George

V
Valacchi, Giuseppe
Vallejo, Jesus
Van Itallie, Christina
van Kooyk, Yvette
Vasta, Gerardo
Vauclair, Jacques
Vestweber, Dietmar
Villa, Anna
Vina, José
Vogt, Brent
Vucenik, Ivana

W
Walker, Edward
Wardemann, Hedda
Wareham, Nick
Warth, Richard
Washburn, Loraine
Watson, Alastair
Weber, Christopher
Weiner, Howard
Weinkove, Ben
Wenner, Peter
Whelan, Christopher
Wherrett, Diane
Wichmann, Thomas
Wild, Martin
Wiley, Jenny
Wilkie, David
Williams, Ernest
Wilson, Patrick
Witten, Ilana
Witting, Larz
Wolfe, Gil
Wood, Susan
Woods, James
Wu, Jian-Young

X
Xu, Weili

Y
Yamamoto, Takamitsu
Yamanashi, Yuji
Yamashita, Yukiko
Yoon, Eunice
Young, Jared
Young, John
Young, Kevin
Yu, Alan
Yuen, Tony

Z
Zaidi, Mone
Zallone, Alberta
Zetter, Bruce
Zhang, Weiqi
Zilles, Karl
Ziskind-Conhaim, Lea
Zon, Leonard